Neoproterozoic Glacial and Associated Facies in the Tanafjord-Varangerfjord Area, Finnmark, North Norway

by

A.H.N. Rice
Department of Geodynamics and Sedimentology
University of Vienna
Althanstrasse 14
1090 Vienna
Austria

Marc B. Edwards
3420 Yoakum Blvd
Houston, Texas 77006
USA

T.A. Hansen
Talisman Energy Norge AS
Verven 4
PO Box 649
Stavanger
Norway

THE
GEOLOGICAL
SOCIETY
OF AMERICA®

Field Guide 26

3300 Penrose Place, P.O. Box 9140 ▪ Boulder, Colorado 80301-9140, USA

2012

Copyright © 2012, The Geological Society of America (GSA), Inc. All rights reserved. Copyright is not claimed on content prepared wholly by U.S. government employees within the scope of their employment. Individual scientists are hereby granted permission, without fees or further requests to GSA, to use a single figure, a single table, and/or a brief paragraph of text in other subsequent works and to make unlimited photocopies of items in this volume for noncommercial use in classrooms to further education and science. Permission is also granted to authors to post the abstracts only of their articles on their own or their organization's Web site providing the posting cites the GSA publication in which the material appears and the citation includes the address line: "Geological Society of America, P.O. Box 9140, Boulder, CO 80301-9140 USA (http://www.geosociety.org)," and also providing the abstract as posted is identical to that which appears in the GSA publication. In addition, an author has the right to use his or her article or a portion of the article in a thesis or dissertation without requesting permission from GSA, provided the bibliographic citation and the GSA copyright credit line are given on the appropriate pages. For any other form of capture, reproduction, and/or distribution of any item in this volume by any means, contact Permissions, GSA, 3300 Penrose Place, P.O. Box 9140, Boulder, Colorado 80301-9140, USA; fax +1-303-357-1073; editing@geosociety.org. GSA provides this and other forums for the presentation of diverse opinions and positions by scientists worldwide, regardless of their race, citizenship, gender, religion, sexual orientation, or political viewpoint. Opinions presented in this publication do not reflect official positions of the Society.

All photos courtesy of Hugh Rice except for Figures 44B and 65B.

Published by The Geological Society of America, Inc.
3300 Penrose Place, P.O. Box 9140, Boulder, Colorado 80301-9140, USA
www.geosociety.org

Printed in U.S.A.

GSA Books Science Editors: Kent Condie and F. Edwin Harvey

Library of Congress Cataloging-in-Publication Data

Rice, A. H. N. (A. Hugh N.)
 Neoproterozoic glacial and associated facies in the Tanafjord-Varangerfjord area, Finnmark, north Norway / by A.H.N. Rice, Marc B. Edwards, T.A. Hansen.
   p. cm.
 Includes bibliographical references.
 ISBN 978-0-8137-0026-7 (pbk.)
 1. Facies (Geology)—Norway—Tanafjorden. 2. Facies (Geology)—Varanger Fjord (Norway and Russia) 3. Geology—Norway—Finnmark fylke. 4. Geology, Stratigraphic—Proterozoic. I. Edwards, Marc B. II. Hansen, T. A. (Tor Arne) III. Title.
 QE281.R53 2012
 554.84'6—dc23

2011053489

Front cover: Early evening view from Geaidnonjoasèohkat, northeastward to the peninsula (Álddarbákti) between Máldovuotna in the west and Vestertana in the east and thence across Tanafjord to Lille Molvika, on Varangerhalvøya, in the east (see Figs. 1 and 3 for direction of view). The hills on Varangerhalvøya are mostly quartzitic sandstones of the pre–Smalfjord Formation (≡Marinoan Glaciation) Tanafjord Group. The rounded steep-sided hills on the photo's west side are of the Digermul Group, reaching from the Cambrian up to the Tremadocian. The low-lying rugged areas between, on the west side of Tanafjord and around Vestertana, are from the Vestertana Group (Ediacaran-Cambrian), here mostly of the Nyborg Formation, Mortensnes Formation (≡Gaskiers Glaciation), and younger rocks. The picture was taken standing on the Thin Sub-Member of the Middle Member of the Mortensnes Formation. The Thick Sub-Member forms the long buff-colored outcrop (Hárjas) beyond Louvdajavrit (the large lakes) on the southeast side of Máldovuotna. The ridges in the foreground, around and above the small lake, reflect the periodic mesoscale east-vergent folding in Member B of the Nyborg Formation.

10 9 8 7 6 5 4 3 2 1

Contents

Abstract 1

Introduction 2

Logistics 2
    Excursion timing 2
    Travel arrangements 2
    Accommodations 2
    Field logistics 3
    Maps and literature 3

Geological Introduction 4
    Stratigraphy, sedimentology, and paleontology 6
    Structure and metamorphism 8
    Isotopic age constraints 8
    Magmatic history 9
    Paleolatitude and paleogeography 9
    Glacigenic deposits and associated strata 9
        Smalfjord Formation (≡ Marinoan Glaciation; late Cryogenian) ... 10
        Nyborg Formation (interglacial succession) 11
        Mortensnes Formation (≡ Gaskiers Glaciation) 12
        Innerelv Member (postglacial) 12
        Facies and paleogeographic interpretation 12
    Stable isotopes and Neoproterozoic correlations 14

Natural History and Cultural Background 15

Excursion Route and Road Log 19
    Day 1. Lower Smalfjord Formation in South Varangerfjord and West Karlebotn,
    Varanger paleovalley (I) 19
        Introduction 19
        Stop 1.1: Lattanjar'ga Peninsula 19
        Stop 1.2: Gåppebalta and Sirdagåp'pi, S. Karlebotn 21
        Stop 1.3: Mar'kan (Karlebotn village) 23
        Stop 1.4: Karlebotn School 24
    Day 2. Lower Smalfjord Formation at Oaibaččanjar'ga and Selešnjar'ga,
    Varanger paleovalley (II) 24
        Introduction 24
        Stop 2.1: Vesterelv 25
        Stop 2.2: Oaibaččannjar'ga (Bigganjar'ga, Cape Headache) 25
        Stop 2.3: Larajæg'gi and unnamed hill 30

Contents

Abstract 1

Introduction 2

Logistics 2
   Excursion timing 2
   Travel arrangements 2
   Accommodations 2
   Field logistics 3
   Maps and literature 3

Geological Introduction 4
   Stratigraphy, sedimentology, and paleontology 6
   Structure and metamorphism 8
   Isotopic age constraints 8
   Magmatic history 9
   Paleolatitude and paleogeography 9
   Glacigenic deposits and associated strata 9
      Smalfjord Formation (≡ Marinoan Glaciation; late Cryogenian). 10
      Nyborg Formation (interglacial succession) 11
      Mortensnes Formation (≡ Gaskiers Glaciation) 12
      Innerelv Member (postglacial) 12
      Facies and paleogeographic interpretation. 12
   Stable isotopes and Neoproterozoic correlations 14

Natural History and Cultural Background 15

Excursion Route and Road Log 19
   Day 1. Lower Smalfjord Formation in South Varangerfjord and West Karlebotn,
   Varanger paleovalley (I) 19
      Introduction 19
      Stop 1.1: Lattanjar'ga Peninsula 19
      Stop 1.2: Gåppebalta and Sirdagåp'pi, S. Karlebotn 21
      Stop 1.3: Mar'kan (Karlebotn village) 23
      Stop 1.4: Karlebotn School 24
   Day 2. Lower Smalfjord Formation at Oaibaččanjar'ga and Selešnjar'ga,
   Varanger paleovalley (II) 24
      Introduction 24
      Stop 2.1: Vesterelv 25
      Stop 2.2: Oaibaččannjar'ga (Bigganjar'ga, Cape Headache) 25
      Stop 2.3: Larajæg'gi and unnamed hill. 30

Day 3. Lower Smalfjord Formation on Skjåholmen and Vieranjar'ga, Varanger paleovalley (III) 31
    Introduction 31
    Stop 3.1: Southeast coast of Skjåholmen viewed from the boat 31
    Stop 3.2: Skjåholmen eastern end 33
    Stop 3.3: View of NE corner of Vieranjar'ga peninsula 34
    Stop 3.4: Vieranjar'ga peninsula 34

Day 4. Lower Smalfjord Formation and cap dolostones along N. Varangerfjord and Leirpollen, Varanger paleovalley (IV) 40
    Introduction 40
    Stop 4.1: Handelsneset 44
    Stop 4.2: Hammarnes Quarry 46
    Stop 4.3: Gis'kananjåkka (east of Nesseby) 47
    Stop 4.4: View to Ruok'sadas 49
    Stop 4.5: Leirpollen 49

Day 5. Glacigenic rocks in the South Tanafjord-Vestertana area 51
    Introduction 51
    Stop 5.1: Šæresgied'djav'ri 51
    Stop 5.2: Auskarnes 52
    Stop 5.3: Torhop 52
    Stop 5.4: SW Duoivejeakkečohkat (east side Vestertana) 52
    Stop 5.5: Šuoššjåkka 54
    Stop 5.6: East of Poas'tagurra 54
    Stop 5.7: Geaidnonjoasèohkat 54

Day 6. Upper Smalfjord Formation on Gæssenjar'ga 55
    Introduction 55
    Stop 6.1: View of Gæssenjar'ga from the road corner west of Torhop 55
    Stop 6.2: Gæssenjar'ga 57

Day 7. Mortensnes Formation in the Las'sasuolo-Låk'sunjåkka coastal profile, West Tanafjord 63
    Introduction 63
    Stop 7.1: Las'sasuolo-Låk'sunjåkka (Stappugied'di) coastal profile 63

Day 8. Associated Neoproterozoic strata along the Varanger Coast (Vadsø to Hamningberg) 76
    Introduction 76
    Stop 8.1: Sjåbuselva, east of Kiby 76
    Stop 8.2: Ekkerøya bird colony 77
    Stop 8.3: West of Komagnes 78
    Stop 8.4: SSW of Hestmannes 79
    Stop 8.5: Persfjord 79
    Stop 8.6: Storflognakken 79

Acknowledgments 80

References Cited 80

With thanks for sponsoring the vehicles during the excursion in 2008, which made this field guide possible.

With thanks for paying the travel and accommodation costs for Marc Edwards in 2007 and 2008.

With thanks for defraying the costs of the color printing.

Neoproterozoic Glacial and Associated Facies in the Tanafjord-Varangerfjord Area, Finnmark, North Norway

A.H.N. Rice*
Department of Geodynamics and Sedimentology, University of Vienna, Althanstrasse 14, 1090 Vienna, Austria

Marc B. Edwards[†]
Statoil Gulf of Mexico, 2103 CityWest Boulevard, Suite 800, Houston, Texas 77042, USA

T.A. Hansen
Talisman Energy Norge AS, Verven 4, P.O. Box 649, Stavanger, Norway

ABSTRACT

This excursion guide describes the glacigenic rocks of the Smalfjord and Mortensnes Formations, and parts of the intervening Nyborg Formation, at the base of the Vestertana Group in the Tanafjord-Varangerfjord region of East Finnmark, North Norway. These are agreed to be either of glacial (sensu lato) origin or significant to our understanding of Neoproterozoic glacial processes and events. The Smalfjord and Mortensnes Formations have been equated with the Marinoan (636.3 ± 4.9 Ma and 635.5 ± 1.2 Ma) and Gaskiers (584–582 Ma) glaciations, respectively. The rocks, which are superbly exposed, display a wide range of sedimentary lithologies/facies; lodgment, banded, deformation, flow, and melt-out diamictites; as well as glaciomarine, proglacial, and fluvioglacial sediments. Diamictites derived from carbonate units showing the Trezona and Wonoka negative $\delta^{13}C$ anomalies are seen.

Glaciotectonic and/or soft-sediment structures described include folds, normal and thrust faults (imbricate fans), flanking structures, shear-sense criteria (sigma clasts), striations (with paleoloess), lobate-cuspate contacts, downward intrusion of fluidized sediments (with bridge structures), pro- and subglacial channels, nested channels, graded beds, possible iceberg dump structures, ghost clasts, dropstones and lonestones, allochthonous rafts of substrate sediments, ball-and-pillow structures, convolute laminations (pseudonodules), delta foresets, ice-crystal molds, sandstone dikes, and a possible kettle hole. The complex geometry of the sub-Smalfjord unconformity onto older Neoproterozoic sediments and the Archean Baltic Shield is documented at several scales. Marinoan cap dolostones (Maieberg negative $\delta^{13}C$ anomaly), with typical sheet cracking (and "pseudostromatolites") and atypical clastic interbeds, are described from the base of the Nyborg Formation at several localities.

---

*alexander.hugh.rice@univie.ac.at
[†]Current address: 3420 Yoakum Blvd, Houston, Texas 77006, USA.

Rice, A.H.N., Edwards, M.B., and Hansen, T.A., 2012, Neoproterozoic Glacial and Associated Facies in the Tanafjord-Varangerfjord Area, Finnmark, North Norway: Geological Society of America Field Guide 26, 83 p., doi:10.1130/2012.0026. For permission to copy, contact editing@geosociety.org. © 2012 The Geological Society of America. All rights reserved.

An unconformity that might be "hiding" the Sturtian glaciation and dolomite-bearing rocks lying to the north of the Trollfjorden-Komagelva Fault on Varangerhalvøya, thought to be correlatives of the 810 Ma and younger Bitter Springs negative $\delta^{13}C$ event, are also described.

Compared to many other areas with Neoproterozoic glacial sequences, the area is easily accessible and has a wide range of modern facilities.

INTRODUCTION

The excursion was first run as part of the International Geoscience Programme (IGCP) 512 "Neoproterozoic Ice Ages" project during the 33rd International Geological Congress, in Oslo, Norway, 2008. This guide is based on the material used for that excursion, with some comments by participants, both positive and adverse, added here.

Note that this is a *guide* to use in the *field* and is not necessarily fully understandable without actually being in front of the rocks. The aim has been to provide a comprehensive (within reason) coverage of both the typical and the atypical lithologies and sedimentary structures (sedimentary facies) in the area. The guide is not a systematic update on published ideas and interpretations, although this has been done at some outcrops. Readers are strongly encouraged to critically evaluate our interpretations in the field and to publish their own ideas.

LOGISTICS

Excursion Timing

Finnmark lies well north of the Arctic Circle and thus has 24 h of daylight during the midsummer months. The midnight sun disappears around 28 July in the Tana area, although it remains light throughout the night for a week or more afterward. However, nighttime temperatures rapidly drop in August, and a ground frost is not unusual in the first week of that month. Better weather is more likely in late June to early July, and this is the recommended time for an excursion.

Travel Arrangements

The village of Tana Bru (Tana Bridge; Fig. 1) is central to the area considered. The simplest way to reach it is to fly to Kirkenes, in East Finnmark, and pick up prebooked rented vehicles. Ordinary cars or minibuses are sufficient since off-road driving is illegal and is not required for the trip.

For groups driving from Central Europe, the quickest route is to drive to Sweden via Lübeck, in Germany, and the Puttgarden-Rødby and Helsingör-Helsingborg ferries that link Denmark to Germany and Sweden, respectively; book both in advance in order to get priority loading during peak-season traffic. The ferries provide nice breaks from driving.

From Helsingborg, take the motorway to Stockholm and on to Uppsala and then the main coastal road to Haparanda and north, through Rovaniemi, in Finland, to Utsjoki. Cross the bridge into Norway there and turn east to Tana Bru. The whole trip can be done nonstop if several drivers are available. Note that speeding in Scandinavia is treated extremely harshly by the police.

For the days requiring a boat in the Tanafjord area, the Kommune office (local government offices) in Tana Bru (www.tana.kommune.no) can be contacted in advance for ideas about locals who may be willing to do this. Boat rides to Skjåholmen and the east end of Vieranjar'ga can be arranged with Grasbakken Cabin and Boat Rental (www.grasbakken.com), based on the south coast of Varangerfjord. The cost of boat rides varies enormously and cannot be estimated here; some locals will take you for free, but others require payment.

Accommodations

Hotel accommodations are not readily available. There is a hotel building in Tana Bru, but whether it is operating as a hotel in any given year is uncertain; it would be costly to use, anyway. This is similarly true for the hotel in Varangerbotn. Accommodations are available in Vadsø, but this entails considerable extra driving and would be costly.

For organized parties, sleeping on the floor of the school in Seida might be possible; hot showers and simple cooking facilities are available, but there are no beds. For this, the school principal must be contacted, well in advance. Contact details can be obtained from the Kommune office in Tana Bru.

Camping is a simple and cheap alternative. There are several camp sites in the area, but again, these are not always operating and should be checked in advance. However, "wild" camping is allowed anywhere as long as you are more than 150 m from the nearest building; are not on obviously private land, farmland, a lay-by, or a roadside picnic area; and there is no sign specifically forbidding camping. You can stay for up to two nights except in very remote areas, where there is no time length restriction. Asking if you can stay on private land or nearer buildings generally produces a friendly positive response. A high-quality tent, sleeping bag, and proper insulating sleeping mat are absolutely essential. Cutting down trees is illegal, as is having open fires between 15 April and 15 September. Food and auto fuel can be bought somewhat more cheaply at Nuorgam, in Finland, only a few kilometers south of Tana Bru, on the east side of the Tana River (Fig. 1).

Note that, since the accommodation arrangements are unknown, the description for how to reach the outcrops starts each day from an easily recognizable geographic spot.

Subsequent locations for that day are described relative to the previous stop.

Field Logistics

Participants must be fully prepared for possible wet, windy, and cold fieldwork weather. A few parts of the trip can be rather arduous, with some walks across country and several boat rides.

Personal hard hats should be worn at all times at outcrops below or near cliffs. It is not the responsibility of the field trip leaders to ensure this.

Warm clothes, including a hat/cap, waterproof jacket, and leggings *must* be taken when making boat journeys, even on a sunny, wind-free day. It is always colder on the water, and the weather or sea conditions may change very significantly and rapidly during a day. If you do not have leggings, and do not wish to buy any, improvise a pair from thick plastic trash can liners.

Ensure that suitable life vests are provided by boat owners. These must be worn, properly fastened, at all times when in the boats. It is far too late to put them on when an emergency arises.

Note that rocks covered in either seaweed or lichen are extremely slippery when wet. A pair of binoculars will be very useful for many parts of the excursion, especially during the boat rides.

Maps and Literature

Although in need of some reworking to reflect a modern understanding of glacial processes, by far the most comprehensive published description available of the glacial and interglacial rocks is Edwards (1984; a more extensive description is given in the online D.Phil. thesis of Edwards, 1972), which can be obtained from the Norwegian Geological Survey (Norges Geologiske

Figure 1. Map showing the main towns, roads, museums, and settlements in the Tanafjord-Varangerfjord area (although many settlements not on a road have been abandoned). Areas and names/numbers of the 1:50,000 topographic and available geological map sheets covering the regions visited during the excursion are indicated.

Undersøkelse; http://www.ngu.no), as can the regional 1:250,000 and 1:50,000 geological maps of the area. Other useful literature can be obtained from the citations given here. For those wishing to examine a wider range of rocks than given here, the field guide of Siedlecka and Roberts (1992) is recommended.

High-quality, 1:50,000 topographic maps can usually be obtained at gas stations, bookstores, and supermarkets, if they are not bought in advance. Note that the grid referencing systems on the 1:50,000 map editions 2-NOR and 3-NOR are different, and coordinates from one edition are not directly applicable to the other. The topographic and available geological maps that cover the areas visited by the excursion are shown on Figure 1.

The Cappelens Kart 5, *Troms og Finnmark*, widely available, gives an excellent overview of the topography, as well as all roads and settlements of the region.

The spelling of Saami place names is once again being revised. Thus, the names given in this guide may soon be out of date officially, although they are still used on many maps that are for sale; the spelling differences are generally minor.

GEOLOGICAL INTRODUCTION

The area considered lies at three major geological boundaries. Paleogeographically, it lies at the border between the NNE-SSW–trending (all orientations are present day) Neoproterozoic to early Paleozoic continental margin of western Baltica, now exposed in the Caledonian nappes, and the contemporary WNW-ESE–trending Timanian basin, relicts of which are seen sparsely along the northern coast of Norway and Russia (Siedlecka, 1975; Siedlecka et al., 2004; Gayer and Rice, 1989). Structurally, it lies across the boundary between deformation associated with crustal shortening (thrusting) in the Scandinavian Caledonides in the west and the underlying autochthonous Baltic Shield (Gayer et al., 1987). Finally, it also lies at the border of the Scandinavian Caledonides in the west and Timanide deformation in the east (Siedlecka et al., 2004).

The Scandinavian Caledonides have been divided into an Autochthon, Lower Allochthon, Window Allochthon, Middle Allochthon, Upper Allochthon, and Uppermost Allochthon (Fig. 2). The Upper Allochthon is divided into Seve-type nappes and overlying Köli-type nappes.

Rocks including the Seve-type nappes down to the Lower Allochthon (the Baltic cover sheets and crystalline sheets of Hossack and Cooper, 1986) represent the strongly shortened Baltoscandian continental shelf, including the ocean-continent transition, overlying a late Neoproterozoic to lower Paleozoic autochthonous sedimentary sequence resting with a marked unconformity on basement rocks of the Baltic Shield. Of these units, only the Autochthon (East Finnmark Autochthon) and Lower Allochthon (Gaissa Thrust Belt), with a poorly defined parautochthonous intermediary zone, are of importance here (Figs. 2 and 3).

The Uppermost Allochthon was derived from the Laurentian side (western side) of the collision zone, while the Köli-type nappes had an intra-oceanic (Aegir-Iapetus oceans) origin.

In the northeast, the North Varanger Region is an important relict of the Timanian belt (Fig. 2; Siedlecka, 1975, 1985; Siedlecka et al., 2004; Siedlecki, 1980). The stratigraphy and structural history of this area differs from that of the rocks to the southwest, from which it is separated by a major dextral

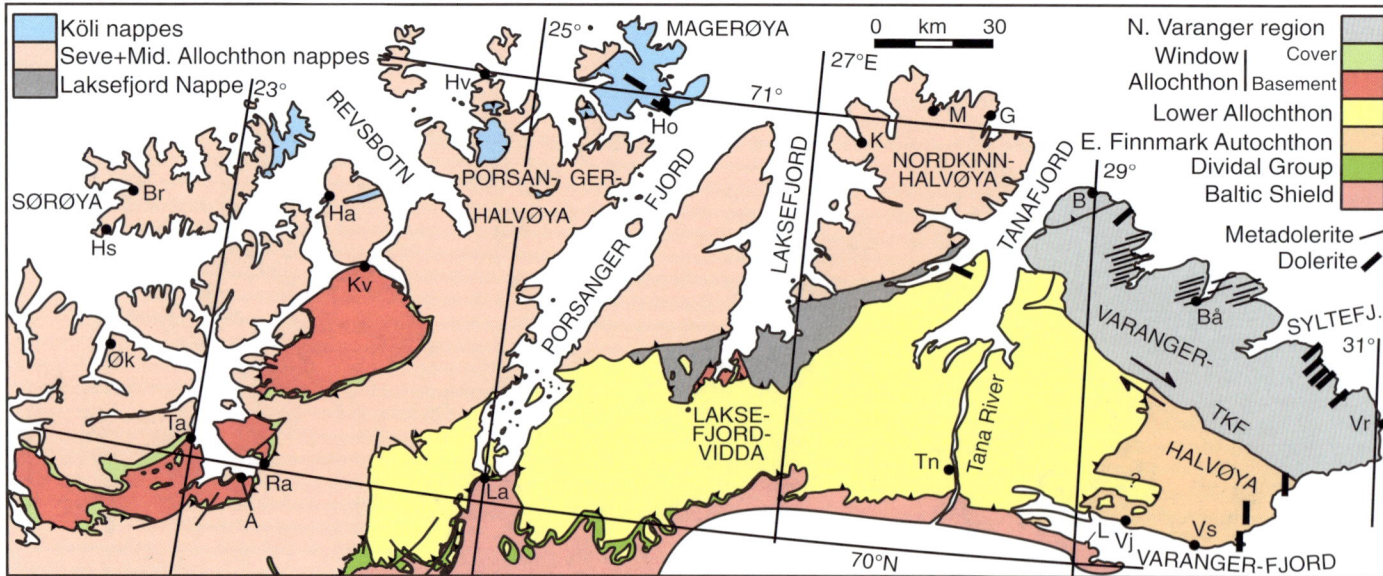

Figure 2. Simplified geological map of the Finnmark Caledonides. All abbreviations refer to towns (except those in italics): A—Alta; B—Berlevåg; Bå—Båtsfjord; Br—Breivik; G—Gamvik; Ha—Hammerfest; Ho—Honningsvåg; Hs—Hasvik; Hv—Havøysund; K—Kjøllefjord; Kv—Kvalsund; L—*Lattanjar'ga*; La—Lakselv; M—Mehamn; Øk—Øksfjord; Ra—Rafsbotn; Ta—Talvik; TKF—*Trollfjorden-Komagelva Fault*; Tn—Tana Bru; Vj—Vestre Jakobselv; Vr—Vardø; Vs—Vadsø.

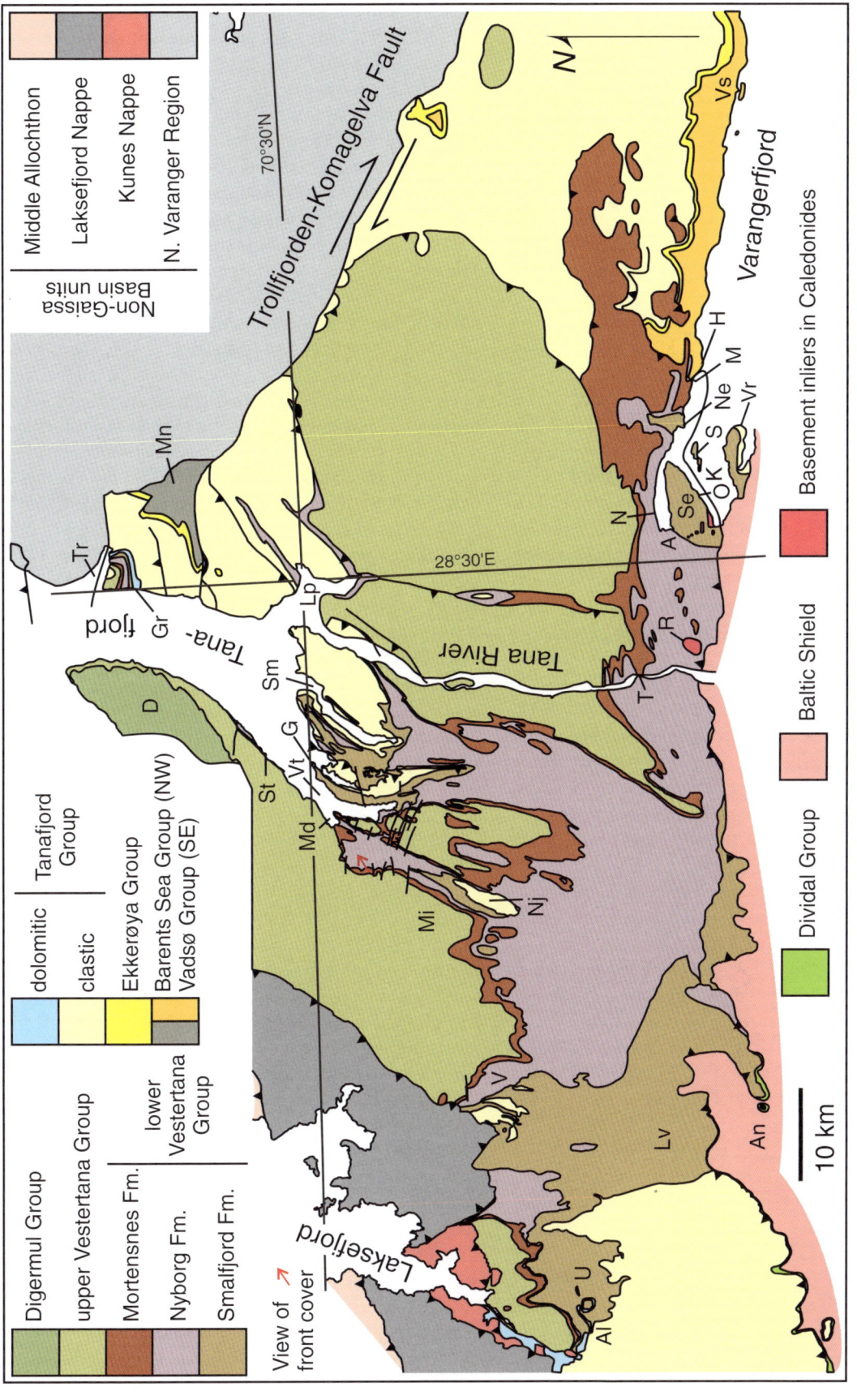

Figure 3. Geological map of East Finnmark. A—Aune; Al—Alduskaidi; An—Andabakoaivi; D—Digermulhalvøya; G—Gæssanjar'ga; Gr—Grasdalen; H—Handelsnes; K—Karlebotn; Lp—Leirpollen; Lv—Laksefjordvidda; M—Mortensnes; Md—Máldovuotna; Mi—Mielkajåkka; Mn—Manjunnas; N—Nyborg; Ne—Nesseby; Nj—Njukčagai'sa; O—Oaibaččannjar'ga; R—Ruos'soai'vi; S—Skjåholm; Se—Selešnjar'ga; Sm—Smalfjord; St—Stappugied'di; T—Tana Bru; Tr—Trollfjord; U—Ukcaskai'di; V—Vaddesbak'ti; Vr—Vieranjar'ga; Vs—Vadsø; Vt—Vestertana.

strike-slip fault, the Trollfjorden-Komagelva Fault (Siedlecki, 1980; Rice et al., 1989a).

Stratigraphy, Sedimentology, and Paleontology

The stratigraphy of the Lower Allochthon–East Finnmark Autochthon consists of five lithostratigraphic groups (Vadsø, Ekkerøya, Tanafjorden, Vestertana, and Digermul Groups; Fig. 4). This is the nomenclature of Rice and Townsend (1996); in other literature (e.g., Johnson et al., 1978; Siedlecka, 1995; Siedlecka and Roberts, 1992; Siedlecka et al., 2004), the Vadsø and Ekkerøya Groups are combined into a single unit, the Vadsø Group, which has an Ekkerøya Formation. All these groups (of Rice and Townsend, 1996) are separated by an unconformity, except between the Vestertana and Digermul Groups. The succession is overwhelmingly siliciclastic; thick dolostones are found only at the top of the Tanafjord Group, and thin dolostones occur at two levels toward the base of the Vestertana Group. The two major glacial events documented in this guide, which rework the aforementioned dolostones, also lie at the base of the Vestertana Group; both are underlain by major unconformities, as is the base of the unit deposited between the glacial events.

The inferred chronostratigraphic ages of these units, which are the tectonically shortened sedimentary fill of the Gaissa Basin, are given in Figure 5; the limited constraints on the chronostratigraphy are given later.

The Vadsø Group is an either 275–470-m- or 575–770-m-thick succession, with five or six formations; the cause of the uncertainty concerns whether the lowest formation (Veinesbotn Formation; 300 m) is a part of the Vadsø Group or of the overlying Tanafjord Group (Røe, 2003). The group primarily crops around Varangerfjord (Røe, 1970, 1987; Banks et al., 1971, 1974; Banks and Røe, 1974; Hobday, 1974), but it is also present in the core of large-scale anticlines near Leirpollen and also in positive flower structures adjacent to the Trollfjorden-Komagelva Fault (Fig. 3). The depositional facies in the Vadsø Group pass from a shallow-marine succession (Veinesbotn Formation) to an alternating sequence of braided fluvial, prodelta and deltaic sands and siltstones (see previous references).

Jensen and Wulff-Pedersen (1996) correlated the Veinesbotn Formation at Oaibaččannjar'ga (Fig. 3) with the Vestertana Group (Fig. 4), but Edwards (1997) and Rice and Hofmann (2000) disputed this strongly. More credibly, Røe (2003) correlated the Veinesbotn Formation with the Gamasfjell Formation, in the Tanafjord Group, although lithological observations by the authors of the Veinesbotn Formation on the north shore of Vieranjar'ga and on Skjåholmen suggest a correlation with the Dakkovarre Formation more likely, if any such correlation is to be made. The only purported direct contact between the Veinesbotn Formation and overlying Klubbnasen Formation is on Skjåholm, where a few meters of fine-grained dark-gray mudstones ascribed to the latter crop out above the former. If the mudstones are indeed from the Klubbnasen Formation, then the correlation of Røe (2003) is invalidated. However, dark-gray mudstones and siltstones also occur within the Veinesbotn Formation, so this is difficult to prove.

The Lattanjar'ga unit, a series of small outcrops of pale-yellow sandstones that crop out on Lattanjar'ga, on the south side of Varangerfjord (Fig. 2), was described as a correlative of the Veinesbotn Formation by Rice et al. (2001), although the Smalfjord Formation is a possible and now preferred alternative (the Lattanjar'ga unit is included in the excursion).

The Lille Molvika Formation (15–190 m; Ekkerøya Group; Johnson, 1978) lies unconformably above the Vadsø Group, except in the Manjunnas area, where it unconformably overlies a combination of the Tyvjofjellet and Båtsfjord Formations of the Barents Sea Group (Siedlecka and Siedlecki, 1971; Rice, 1994; Fig. 5). The Lille Molvika Formation consists of typically dark-gray, but sometimes red- or green-tinged mudstones to siltstones with thinly interbedded white, sometimes rusty stained or spotted, thin, irregularly bedded sandstones of shallow-marine origin. On the basis of acritarch assemblages, Vidal (1981) suggested that there was a marked time gap between the base of the Ekkerøya Group and the top of the Vadsø Group (Fig. 5); this was a major reason for separating the rocks into two groups (Rice and Townsend, 1996). The contact is also a major erosional surface, which might reflect an earlier Cryogenian glaciation (see Stop 8.1).

Siedlecka and Siedlecki (1971) defined seven formations in the unconformably overlying Tanafjord Group, with a total thickness of 1448–2015 m (Fig. 4). Halverson et al. (2005) and Rice et al. (2011) informally divided the uppermost unit, the Grasdal Formation into two formations, the Grasdal and overlying Fadnuvaggi formations. The contact of the Tanafjord Group with the underlying Ekkerøya Group is a very low-angled unconformity with little erosional relief (Johnson, 1978; Rice, 1994).

Sedimentological studies indicate that the bulk of the Tanafjord Group is primarily of shallow-marine clastic origin. Fluctuations in sea level led to minor regressions and transgressions, affecting the precise environment. The lower part of the dolomitic Grasdal Formation has been interpreted broadly as a sabkha deposit, and the upper part has been interpreted as an intertidal sequence, with the source area lying essentially to the southeast. The Grasdal Formation is overlain by the informal Fadnuvaggi formation, made up of ~20 m of dark-gray to black alternating siltstones and sandstones, locally containing pyrite nodules up to several centimeters in size.

The overlying Vestertana Group (Reading and Walker, 1966), consisting of five formations with a thickness of 1110–2025 m, rests with a regionally marked subglacial unconformity on the underlying successions, although at outcrop scale, this might not be apparent. The large variation in thickness is partly due to a major unconformity between two formations within the group, also of subglacial origin, but due to a different glaciation. The basal three formations, which include the glacial deposits, the subject of this excursion, are described in more detail later herein.

The Vestertana Group is conformably overlain by the Digermul Group (Reading, 1965; Figs. 4 and 5), which includes the

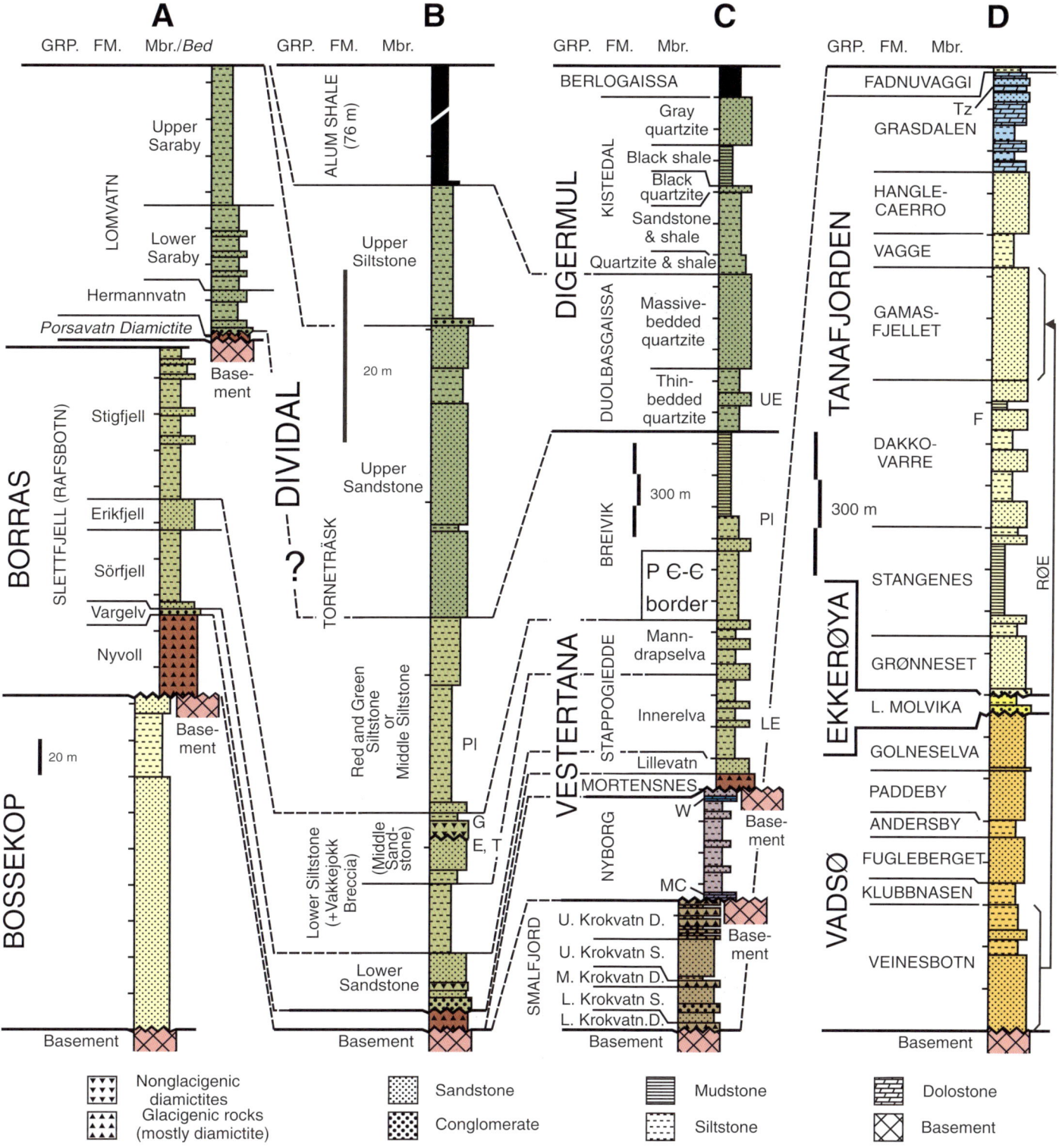

Figure 4. Overview of the stratigraphy of (A) the Komagfjord Antiformal Stack (Window Allochthon); (B) the Dividal Group (autochthon); (C and D) the Gaissa Thrust Belt and East Finnmark Autochthon (Lower Allochthon and Autochthon; see Fig. 2). E—Ediacaran fauna; F—Ferruginous sandstone member; G—?glendonite; Pl—*Platysolenites antiquissimus*; T—*Treptichnus pedum*; UE, LE—uppermost and lowermost reported occurrences of Ediacaran fauna. Tz—Trezona $\delta^{13}C$ anomaly; MC—Marinoan cap-dolostone and Maieberg $\delta^{13}C$ anomaly; W—Wonoka $\delta^{13}C$ anomaly. The $\delta^{13}C$ anomalies are from Halverson et al. (2005), based on worldwide correlations. The 420 m thickness of the Smalfjord Formation on Laksefjordvidda is from Føyn and Siedlecki (1980); D.—diamictite; S.—sandstone. The line RØE refers to the correlation of the Veinesbotn Formation with the Gamasfjell Formation proposed by Røe (2003). Figure is modified after Rice et al. (2011) and Stodt et al. (2011).

Tremadocian Berlogaissa Formation, the youngest stratigraphic unit preserved within the Gaissa Basin.

Acritarchs have been documented from the Vadsø, Ekkerøya, Tanafjorden, and Vestertana Groups, though only reworked acritarchs and vase-shaped fossils have so far been found in the Smalfjord and Mortensnes Formations (Vidal, 1981; Vidal and Siedlecka, 1983). A limited Ediacaran fauna (Farmer et al., 1992; Crimes and McIlroy, 1999) is dominated by discoidal forms, including *Aspidella* (Narbonne, 2008, personal commun.). The oldest forms occur in the Innerelv Member (Fig. 4).

The Cambrian-Precambrian boundary (542 Ma) lies near the base of the Breivik Formation (Fig. 4; Banks, 1970; Føyn and Glaessner, 1979; Farmer et al., 1992; Crimes and McIlroy, 1999). Chronostratigraphic constraints are outlined later.

Structure and Metamorphism

The internal part of the Lower Allochthon, called the Gaissa Thrust Belt in the Porsangerfjord area (Fig. 2), is a strongly imbricated succession stratigraphically correlated with the Tanafjord, Ekkerøya, and Vadsø Groups (Williams, 1976; Townsend et al., 1989; Rice and Townsend, 1996). Balanced cross sections indicate that ~50% shortening occurred, which is typical for external imbricate zones, during E- to ESE-directed brittle thrusting (Townsend et al., 1986, 1989; Gayer et al., 1987).

In the Laksefjordvidda area and eastward, shortening decreases to ~15% (Chapman et al., 1985), and both the orientation (Williams, 1979) and scale of the folds vary. Fold axes near the southern margin of the thrust belt east of Laksefjord trend E-W to ENE-WSW and face south. Northward, the axial trend rapidly swings round to a NNE-SSW to N-S, east-facing, orientation. This persists up to the Vestertana-Leirpollen area, where the orientation swings toward NE-SW, through Digermulhalvøya and the area north of Leirpollen (Fig. 3). Still further north, in the Grasdalen area, immediately south of the Trollfjorden-Komagelva Fault (Fig. 3), fold axes swing back through E-W to ESE-WNW and essentially merge with the south-southwest–facing flower-structure folds adjacent to the Trollfjorden-Komagelva Fault (Siedlecki, 1980; Rice et al., 1989a).

To the east of the Tana River (Fig. 3), folds near the southern margin of the thrust belt swing from an E-W orientation to a WNW-ESE, southward-facing orientation along the northern shore of Varangerfjord. This zone persists as far east as exposures of the Nyborg Formation occur (Fig. 3). Deformation dies out gradually from the Gaissa Thrust Belt into the East Finnmark Autochthon within central Varangerhalvøya.

Metamorphic conditions decrease from upper anchizone in the west of the Gaissa Thrust Belt to diagenetic zone in the East Finnmark Autochthon. The underlying autochthonous Dividal Group also records diagenetic zone metamorphism (Rice et al., 1989b). North of the Trollfjorden-Komagelva Fault, epizone metamorphic conditions occurred (Rice et al., 1989a).

Isotopic Age Constraints

Robust, precise isotopic age constraints for the age of sedimentation in this area are lacking, since no volcaniclastic deposits have been identified. Several workers have undertaken K-Ar, Rb-Sr, and $^{40}Ar/^{39}Ar$ whole-rock and separates studies, but while the results gave broadly Caledonian or Timanian deformation

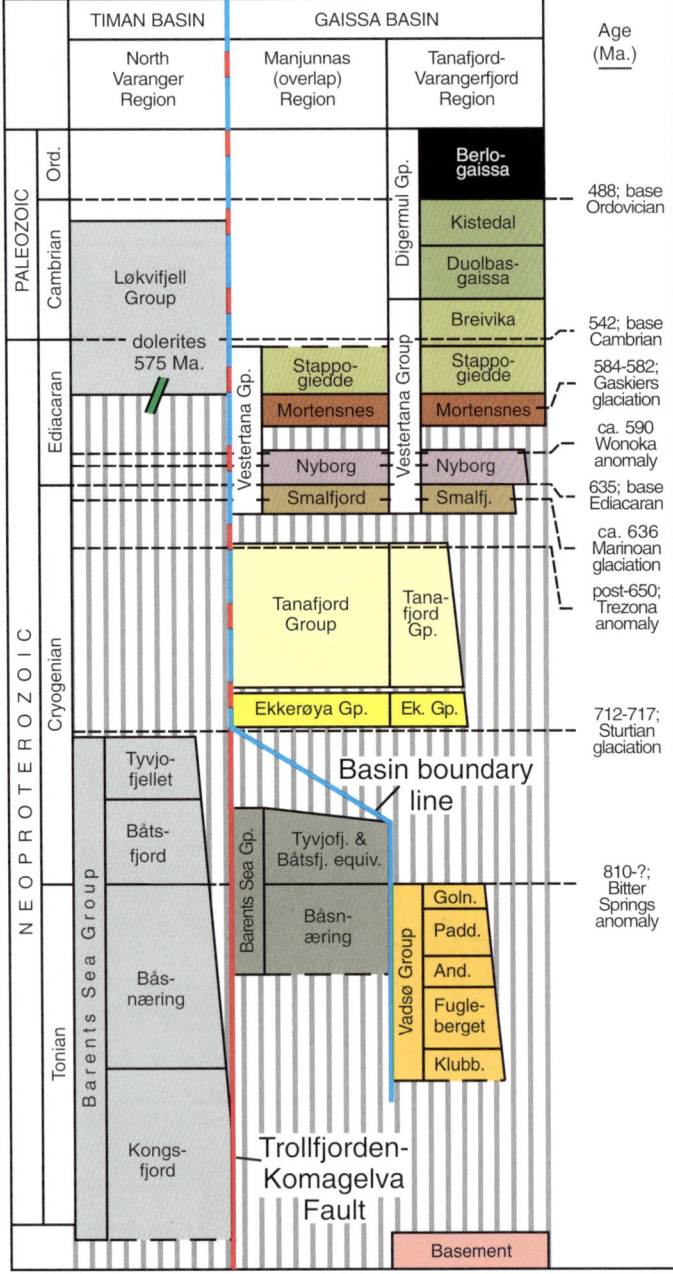

Figure 5. Inferred chronostratigraphy of the Neoproterozoic sediments in East Finnmark, adapted from Siedlecka (1995), with data from Rice et al. (2004) and Halverson et al. (2005). The model of Røe (2003) has been incorporated; hence there is no Veinesbotn Formation (compare with Fig. 4). Abbreviations: Goln—Golneselva, Padd.—Paddeby, And.—Andersby, Klubb.—Klubbnasen.

ages, or ages inferred to indicate the time of diagenesis/deposition, they are only "acceptable" in that they fit with the preconceived geology of the area; anomalous data are rejected, rather than taken to signify an important new constraint. Whether deformation was early Caledonian (Finnmarkian; ca. 490 Ma) or late Caledonian (Scandian; ca. 410 Ma), or part of some intervening event, is unclear, but not important here. Further, the amount of pre-Caledonian (Timanian) deformation in East Finnmark is also uncertain, although much speculated upon (cf. Siedlecka et al., 2004).

Some constraints on the age of sedimentation have been derived from stable isotope studies; these are outlined in a subsequent section.

Magmatic History

Mid-ocean-ridge basalt (MORB)–within-plate basalt (WPB) metadolerites along the northern coast of the North Varanger Region (Fig. 2; Roberts, 1975) were intruded at or after 577 ± 14 Ma (Rice et al., 2004), prior to deformation (Rice and Reiz, 1994). These are not seen during the excursion.

Several unaltered within-plate doleritic dikes have been documented on Varangerhalvøya (Fig. 2; Roberts, 1975; Rice et al., 2004). These trend N-S south of the Trollfjorden-Komagelva Fault and NE-SW north of the fault, where they cut a well-developed slaty cleavage. Beckinsale et al. (1976), Guise and Roberts (2002), and Rice et al. (2004) showed that intrusion occurred at ca. 376 Ma. Similar Devonian-aged dikes occur throughout the northern part of the Kola Peninsula, which lies to the east and southeast of Varangerfjord, cutting the Baltic Shield.

Paleolatitude and Paleogeography

At ca. 750 Ma, a southward-facing Finnmark-Kola region at ~15°S formed the southern margin of Baltica (Hartz and Torsvik, 2002). This region lay adjacent to a roughly E-W–trending rift (Timanian margin), which, to the west, linked with the N-S–trending rift/spreading axis developing between Baltica and Laurentia (Greenland). At 616 Ma, Baltica is thought to have been at polar latitudes (75°S) based on the well-dated Egersund dolerite dikes in southern Norway (Bingen et al., 2005). By ca. 550 Ma, East Finnmark was lying at ~50°S (Cocks and Torsvik, 2005). In contrast, Cawood and Piarevsky (2006) argued that Baltica was not necessarily inverted (southwards-facing) in the Neoproterozoic, but was located either at ~30°N or 30°S at 550 Ma. The purported occurrence of glendonite in the autochthonous shallow-marine Vakkejokk Breccia, in the Torneträsk area (Fig. 4, column B), above *Treptichnus pedum* (Stodt, 1987), implies that either Baltica subsequently moved rapidly poleward or that polar ice became markedly more extensive at that time; there is no evidence for either implication. Further, Stodt et al. (2011) cast doubt on the interpretation of the calcite nodules as glendonite.

Glacigenic Deposits and Associated Strata

Two glacigenic units have been recognized in the area (Figs. 3, 4, 5, and 6), the older Smalfjord Formation and younger Mortensnes Formation, separated by the Nyborg Formation (Føyn, 1937; Reading and Walker, 1966; Edwards, 1984; Rice et al., 2011). Both glacigenic units cut down-section toward the south and directly overlie the crystalline basement. The Nyborg Formation also very locally lies on the basement. Neoproterozoic

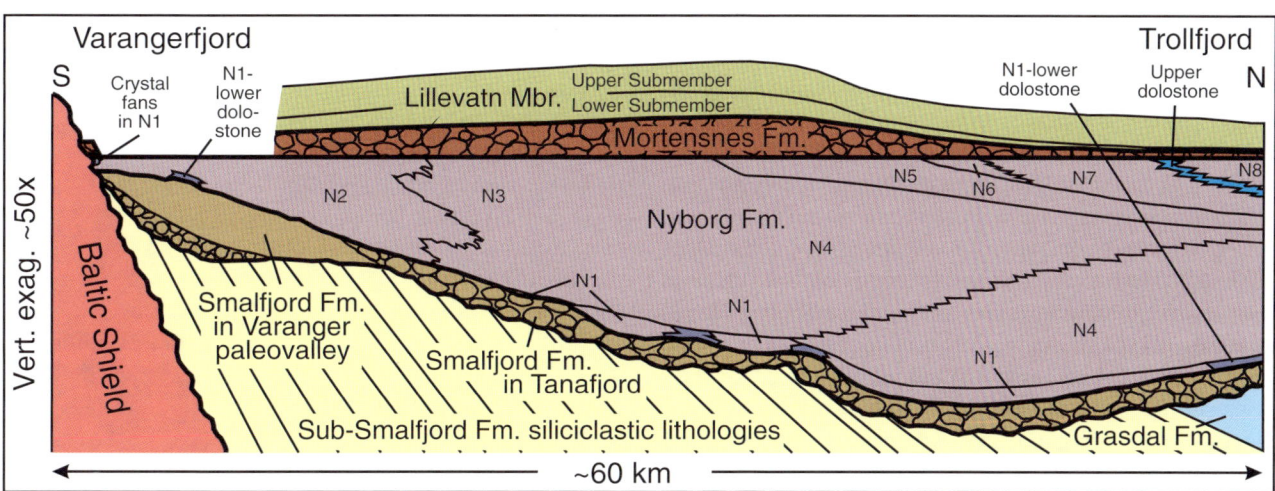

Figure 6. Schematic, essentially N-S–oriented profile from the west end of Varangerfjord to the Trollfjord area of eastern Tanafjord (Fig. 3). This shows the two major subglacial unconformities within the succession, at the base of the Smalfjord Formation, which is the base of the Vestertana Group, and at the base of the Mortensnes Formation. N1–N8 are facies in the Nyborg Formation: N1—cap dolostone; N2—fan channel; N3—submarine fan; N4—basin floor; N5—slope; N6—tidal distributary/bay; N7—lagoon; N8—transgressive barrier/offshore. Profile is drawn with the basal Mortensnes Formation unconformity restored to the horizontal. Figure is modified after Edwards (2004) and Rice et al. (2011).

glacigenic rocks from further west and southwest in the northern Scandinavian Caledonides were reviewed in Stodt et al. (2011); these are all correlated with the Mortensnes Formation.

In the following text, *lonestone* indicates an atypically large, isolated clast within a relatively finer-grained matrix for which there is no evidence that it fell into the sediments, while *dropstone* indicates a clast for which there is evidence that it fell into the surrounding sediments (crosscutting relationships).

Smalfjord Formation (≡ Marinoan Glaciation; Late Cryogenian)

The Smalfjord Formation has been preserved in the Varanger, Tanafjord, and Laksefjordvidda areas (Fig. 3). The latter, which is not included in this excursion, preserves the maximum thickness recorded of ~420 m (Føyn and Siedlecki, 1980), but this is not so well exposed or easily accessible as elsewhere.

In the Varangerfjord region, two diamictite (D1 and D2) and four sandstone-conglomerate (S1–S4) facies have been recognized (Bjørlykke, 1967; Edwards, 1975, 1984; Laajoki, 2001, 2002; Arnaud and Eyles, 2002; Arnaud, 2008; Baarli et al., 2006). D1 is a massive matrix-supported pinkish diamictite with a medium/coarse-grained matrix. The few granitic/sandstone clasts are subrounded to subangular and up to 20 cm across. D1 units can be traced for several hundred meters, in beds <1 m thick. D2, which is typically matrix supported and massive, consists of a gray, muddy, medium-grained sandstone matrix with numerous <50-cm-size clasts of granite and sandstone, some faceted.

S1 is composed of well-stratified pebble conglomerates, erosively based and lenticular, and occasionally showing clast imbrication, cross-bedded and parallel-laminated sandstones, with fining-upward trends between these lithologies. Rare mud drapes occur. S2 is typified by large-scale foresets, 5–10 m thick, laterally traceable for up to several kilometers. These include steeply dipping, poorly sorted, pebble conglomerates to well-sorted, gently dipping, parallel-laminated sandstones. S3 consists of medium-bedded, parallel-sided to slightly lenticular sandstones, usually internally massive, but at times parallel-laminated or rippled. Mudstone partings are thin or absent, but convolute bedding, recumbent folds, and synsedimentary faulting occur. Isolated steep-sided channels, with a poorly sorted conglomerate infill, have been documented. S4 is made up of interbeddded sandstones and mudstones; the former are of thin to medium thickness, laterally continuous beds and with turbiditic features (erosive surfaces, sole marks, grading, convolute bedding). Rare dropstones occur. A 3-m-thick, crudely-stratified sedimentary breccia, sometimes graded, within S4 contains angular sandstone clasts, up to 40 cm across, in a coarse-grained sandstone matrix. Unit X, which has a very limited occurrence at the eastern end of Vieranjar'ga (Edwards, 1975) is similar in facies to S4, but it underlies S1.

Similarly, facies C1 has an even more restricted outcrop on Vieranjar'ga (Edwards, 1975). This consists of a monomict orthoconglomerate with pebbles identical in composition and size to facies D1, but it is clast supported rather than matrix supported. This formed by winnowing of facies D1, with little transport of the coarse fraction prior to the deposition of facies S1. The sediments were deposited by reworking of facies D1 in an ice-marginal or proglacial stream environment and are possibly a channel deposit.

Lithofacies have been traced for several tens of meters laterally, although abrupt facies changes also occur (Edwards, 1984). The diamictites, breccias, and S4 facies tend to occur near the base, whereas the S1 facies occurs near or at the top of the succession.

In the area around Tanafjord (Fig. 3), a completely different lithostratigraphy has been recorded. Based predominantly on outcrops south of Tanafjord, Edwards (1984) recognized five cycles (Members a–e; Figs. 7 and 8) consisting of erosion

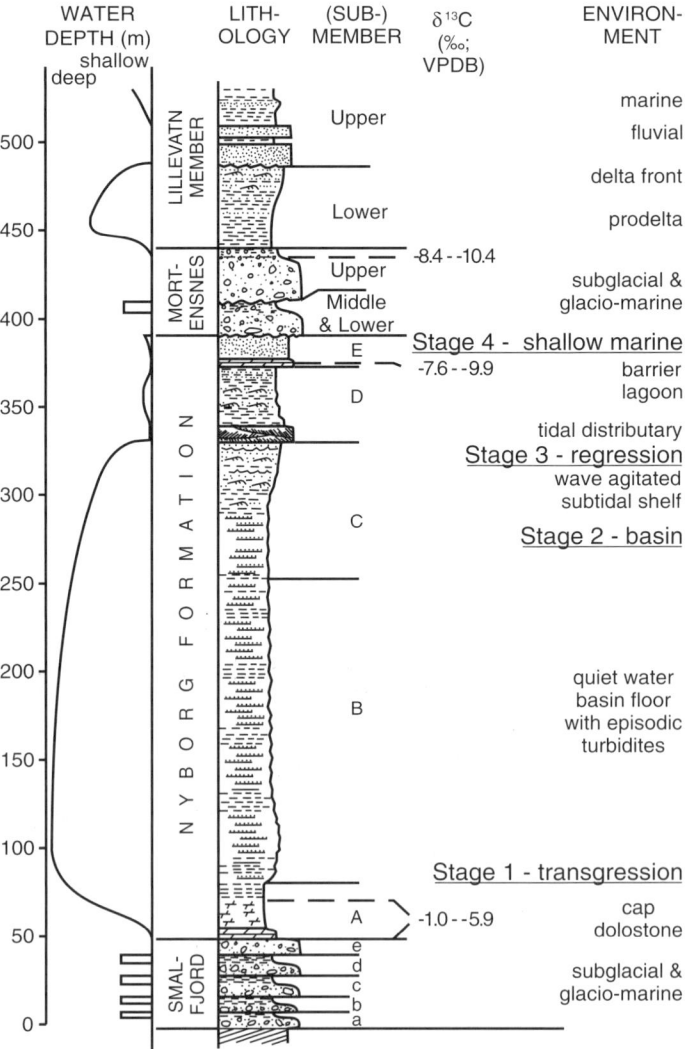

Figure 7. Detailed lithostratigraphy, $\delta^{13}C$ values of dolostones, and inferred water depths and depositional environments of the lower part of the Vestertana Group, after Edwards (1984), with data from Halverson et al. (2005) and Rice and Halverson (2005–2010, personal observations) added. VPDB—Vienna Peedee belemnite.

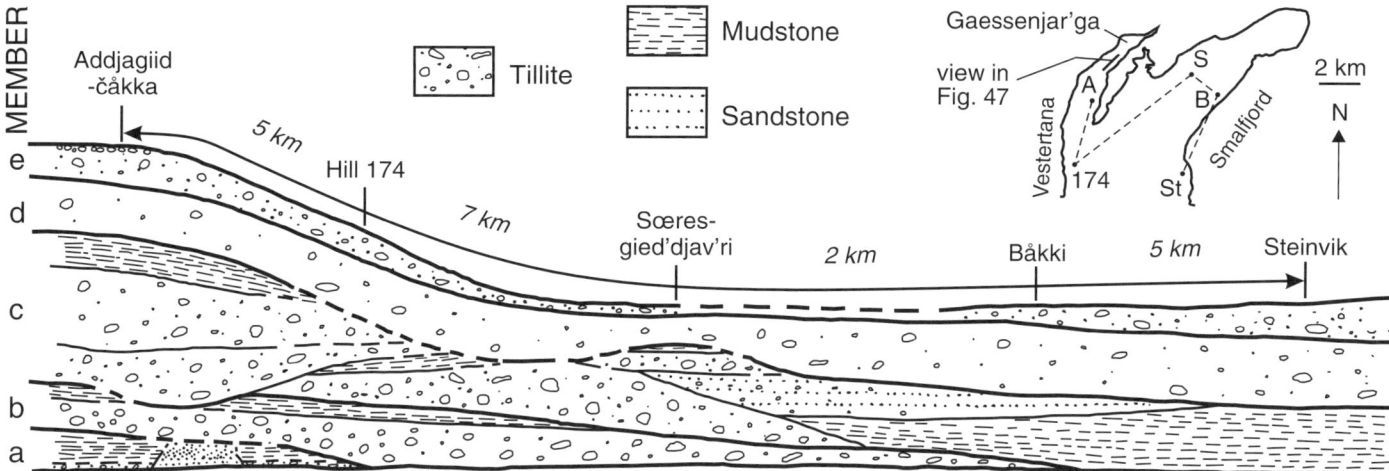

Figure 8. Distribution of the five cyclical depositional members (a–e; cf. Fig. 7) of the Smalfjord Formation in the Vestertana-Smalfjord area (after Edwards, 1984).

surface, diamictites, and laminated mudstones, sometimes with intervening sandstones, in the upper part of the Smalfjord Formation. In part, these members can be traced over hundreds of square kilometers. Diamictites, which are 2–40 m thick, structureless, or stratified and have erosional bases, both regionally and locally, are matrix supported, with <1-m-sized clasts (most <35 cm) in a siltstone to sandstone matrix.

Material from underlying beds was often deformed by the passage of the overlying glacier, forming a basal deformation diamictite; these are made up overwhelmingly of the immediate footwall lithology. The diamictites above this may be an admixture of far-traveled, basement-derived crystalline and sedimentary rocks and more locally derived sedimentary material, forming banded diamictites, with glaciotectonic structures such as folds, faults, boudins, shear bands, load casts, and flame structures. Some diamictite units have gradational contacts (Hansen, 1992; Arnaud and Eyles, 2002). Stratified sandstone/conglomerate bodies, resembling the Varangerfjord facies, occur in the diamictite; these are up to several meters thick and several kilometers long (Fig. 8). Dominant clast and matrix sources determine the diamictite lithology; buff/brown colors indicate a dolostone source (Members b, c, and e; Figs. 7 and 8), most likely the Grasdalen Formation (Figs. 4 and 6). Member a is purple, with a hematitic matrix, derived from ferruginous sandstones of the Dakkovarre Formation (Fig. 4); this is exposed in the Tanafjord Group immediately below the sub-Smalfjord Formation unconformity in this area. Diamictites may also be distinguished by variable clast abundances. Some diamictite units exhibit normal grading with a sandy matrix at the base and a muddy matrix at the top (Hansen, 1992; Arnaud and Eyles, 2002).

Sandstone beds in the Tanafjord area are massive, graded (normal or reverse), rippled, and trough/planar cross-bedded. Laminated mudstones, 0.3–10 m thick, vary in color, lamination prominence, whether the lamination thickness is irregular or rhythmic, and also the size, abundance, and composition of lonestones. The latter are predominantly dolostones, <1–30 cm in size, and are either dispersed throughout the lithology or lie in discrete layers. Some laminated mudstone units are interbedded with sandstones or diamictites.

In the Gæssenjar'ga area, where >85% of the section consists of diamictite, Hansen (1992) and Arnaud and Eyles (2002) recognized essentially the same lithologies as Edwards (1984) but ordered them with less depositional cyclicity.

Southward from Grasdalen (Fig. 3), the unconformity at the base of the Smalfjord Formation cuts down through the 2.5 km thickness of the Tanafjord, Ekkerøya, and Vadsø Groups to the basement around western Varangerfjord (Karlebotn). Minor outliers of Neoproterozoic sediments overlie the basement in this area (Fig. 3), preserving an irregular unconformity, locally paleo-frost-shattered (Bjørlykke, 1967; Edwards, 1984, 1975; Siedlecka, 1990; Rice and Hofmann, 2000; Rice et al., 2001; Laajoki, 2001, 2003). Rare, but spectacular, E-W– to WNW-ESE–oriented subglacial striations have been recorded at the base of the Smalfjord Formation in the Karlebotn and Handelsneset areas (Fig. 3; Reusch, 1891; Bjørlykke, 1967; Rice and Hofmann, 2000; Laajoki, 2002; Bestmann et al., 2006).

On Laksefjordvidda (Fig. 3), Føyn and Siedlecki (1980) identified a 420-m-thick succession with three main glacial successions, the Lower, Middle, and Upper Krokvatn Diamictites, and two interglacial successions, the Lower and Upper Krokvatn Sandstones (Fig. 4). The Lower Krokvatn Diamictite is thought to be equivalent to the Varangerfjord succession, while the Upper Krokvatn Diamictite is thought to be equivalent to the Tanafjord succession. The Lower and Middle Krokvatn diamictites contain basement-derived clasts, while the Upper Krokvatn Diamictite has abundant dolomite clasts.

Nyborg Formation (Interglacial Succession)

The two glacial successions are separated by the 0–370-m-thick Nyborg Formation (Reading and Walker, 1966), with five members (Figs. 6 and 7). The basal Member A, which generally is less than a few tens of meters thick and more clastic upwards, consists of lateral and vertical gradations between (1) massive buff-weathering dolomicrites (facies 1 of Edwards, 1984; here termed facies NA1), sometimes with sheet cracks, filled with isopachous cements, and pseudo-tepees, to (2) interbedded red sandstones/shales and dolomicrites (facies NA2), the latter sometimes with a pinkish hue, and locally reworked as edgewise breccias (facies NA4), to (3) dolomite-cemented orange-red sandstones, to (4) white-weathering sandstones (facies NA5) with interbedded dark-gray shales (Edwards, 1984; Reading and Walker, 1966; Rice and Hofmann, 2001). These facies are listed in their typical order of deposition, but exceptions occur.

Facies NA3 is composed of purple mudstones up to 25 m thick; these typically occur between facies NA2 and Member B. On Ruos'soai'vi, west of Varangerfjord (Fig. 3), 10 mm sulfate crystal fans, now replaced by quartz, grew on the basement, preserved within a 12-mm-thick biotite extraclastic recrystallized dolostone, overlain by pale sandstones (Rice, 1996, personal observation). Similar white-weathering sandstones, probably also of facies NA5 and likely derived from reworking the Smalfjord Formation deposits, directly overlie the very irregular basement surface on the north and northeast slopes of Ruos'soai'vi. Pale-gray– to white-weathering, bluish sandstones also unconformably overlie sheet-cracked dolomicrite in the Mielkajokka and Njukčagai'sa areas (Fig. 3), and these are taken to be facies NA5 as well.

Members B–D of the Nyborg Formation consist of, respectively, interbedded reddish-purple shales/sandstones (~200 m), interbedded gray-green shales/sandstones (~150–200 m), and purple sandstone/gray-green shales (~70 m; Edwards, 1984; Fig. 7). Member E (~25 m thick), which is exposed only in the Trollfjord-Grasdalen area (Fig. 3), consists of white to gray sandstones with thin buff dolomicrites near the base (Edwards, 1984).

The base of the Nyborg Formation is an unconformity resting on different parts of the Smalfjord Formation and, in the area west of Karlebotn, directly on the basement (Føyn and Siedlecki, 1980; Edwards, 1984; Siedlecka, 1990; Rice, 1996, personal observation).

Mortensnes Formation (≡ Gaskiers Glaciation)

Edwards (1984) recognized three members in the <50-m-thick Mortensnes Formation (Fig. 9). The lowest (<30 m) is a northward-thinning wedge of predominantly gray-green to purple massive diamictite of highly variable clast/matrix composition, with basement clasts and minor intrabasinal clasts from the underlying substrate. Brecciation of the substrate occurs at the base of the diamictite and is overlain by displaced substrate blocks up to 20 m long and 1 m thick. Lenses and bands of dark diamictite with extrabasinal clasts (basement-derived) occur, as well as diamictites that have a broader spectrum of clast lithologies. Some clasts may show facets and striations (Banks et al., 1971), while clast sizes decrease overall to the north (Edwards, 1984). The Middle Member has a gradational contact to the underlying rocks and is distinguished by its buff-brown weathering, dolomitic matrix, and clast composition, with subordinate chert clasts. Around the southern end of Tanafjord, the Lower Member thins out entirely, while the Middle Member increases in thickness from south to north, with two submembers. The Thin Submember is made up of 2–4 m of stratified diamictite with primary and soft-sediment deformation structures. The Thick Submember consists of a range of lithologies, including a blanket of massive purple to gray-green deformation diamictite, a zone of large tabular blocks derived from the Nyborg Formation, diamictite, white sandstones, relatively rare stratified dolomitic diamictite, prominent buff-brown diamictite with a sandy matrix that thins from 20 to 8 m from south to north, and, at the top, a bedded buff-brown diamictite. The unconformably overlying Upper Member consists of dark-gray massive diamictite with basement clasts, locally with a thin dolomitic diamictite at the top. Polymict conglomerates overlie much of the formation (Edwards, 1984).

The base of the Mortensnes Formation is a planar surface cutting down-section toward the south at a very low angle. West of Karlebotn, it may lie directly on the basement; on Ruos'soai'vi (Fig. 3), red diamictite consisting mostly of basement clasts lie topographically above Nyborg Formation carbonates (but not in direct contact), suggesting it is younger (Rice, 1996, personal observations). However, similar diamictites also lie topographically below the Nyborg Formation in this area (Siedlecka, 1990; Rice, 1996, personal observations). The contact often exhibits brecciation/homogenization of the underlying sediments (deformation diamictite; Edwards, 1984).

Innerelv Member (Postglacial)

Diamictites of the Mortensnes Formation are typically sharply overlain by laminated mudstones of the Lillevatn Member of the Stappogiedde Formation (40–110 m; Figs. 4 and 7; Reading and Walker, 1966; Edwards, 1984). This member rapidly thins northward, in the vicinity of Vestertana (Figs. 3 and 6; Edwards, 1984). At one road cut, the contact has a thin transitional zone. The Lower Submember (3–55 m) consists of gray, parallel-laminated mudstones, silty to sandy in the north, grading upward into a siltstone, with some ripple cross-lamination and with fine- to medium-grained lenticular sandstones (Edwards, 1984). The Upper Submember is a complex assemblage of sandstone and shale facies, including poorly cross-bedded, coarse arkosic sandstones; granule conglomerates; medium-grained subarkosic sandstones; relatively well-sorted and well-rounded sandstones; thin- to medium-bedded fine- to very fine-grained lenticular, erosive-based sandstones; dark-gray, brown-weathering, rippled and finely laminated micaceous silty-sandy mudstone, sometimes in coarsening-upward cycles; and finely parallel-laminated gray mudstones (Edwards, 1984).

Facies and Paleogeographic Interpretation

Bjørlykke (1967) and Edwards (1984) interpreted the Varangerfjord succession of the Smalfjord Formation as an infill of a glacially scoured northwestward-plunging valley

(Varangerfjord paleovalley; see also Laajoki, 2002; Baarli et al., 2006; note that a glacially scoured valley having an at least partially marine infill should properly be termed a fjord). This paleogeographic interpretation was based on the sharp irregular angular unconformities that can be seen at the base of the formation around Varangerfjord, especially on the north side. Irregular glacier retreat left ice-cored moraines that were submerged by rising water levels. Bathymetric lows were filled first by sediment gravity flows and overlain subsequently by rapidly prograding deltas and sandur plains. Later diamictites may represent glacial advances and/or sediment slumping.

Edwards (1984) interpreted the Smalfjord Formation around Tanafjord as a cyclic deposit of glacial advances and retreats, with a basal massive lodgment tillite, locally with sandstones deposited in situ by subglacial meltwater at the ice margin. Ice movement resulted in the formation of basal deformation tillites of deformed substrate sediments, while mixing of local and far-traveled material formed banded tillites (i.e., tillite with layers of different color/clast composition, reflecting incomplete mixing of material from different provenances; Edwards, 1972). Erosional bases are indicated by deformation structures and mixing with subjacent material. During retreat, poorly sorted sands/silts with pebbles were deposited at the ice margin by tractional underflows and gravity flows. In interglacial periods, finer-grained sediments accumulated, occasionally with dropstones.

Hansen (1992), while agreeing that the diamictites were of mostly glacial origin, argued for a shelf depositional environment, dominated by undermelt and subaqueous mass-movement tillites (flow tillites), some heavily deformed by an advancing glacier. Rapid vertical and lateral thickness variations and terminations of both diamictites and associated facies were mostly associated with changes in glacial movement directions and local bathymetrical conditions rather than glacial erosion and lodgment. Glacigenic flow-tillites laterally grading into sandstones, and, at other stratigraphic levels, grading into rhythmites, indicate transport directions and the relative location of the basin/continent.

In contrast, Arnaud and Eyles (2002) suggested that the diamictites in both the Gæssenjar'ga area (south of Tanafjord) and along Varangerfjord accumulated from subaqueous gravity flows from the basin margin, with the latter forming part of a debris apron (see also Crowell, 1999; Schermerhorn, 1974;

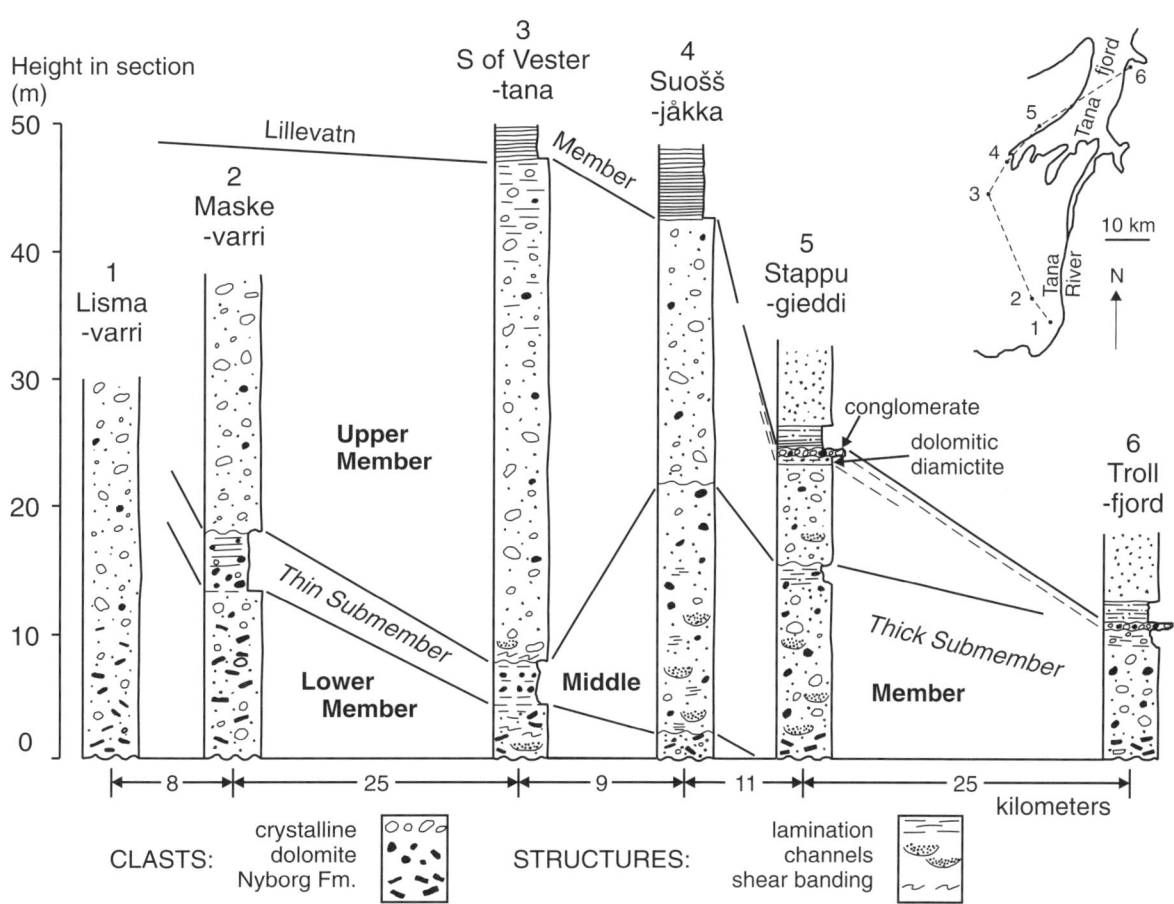

Figure 9. Distribution of the three members within the Mortensnes Formation in the Tanafjord–Tana River area. Note the northward thinning of both the Lower and Upper Members and the concomitant thickening of the Middle Member (consisting of a southerly Thin Submember and a northerly Thick Submember; modified after Edwards, 2004).

Jensen and Wulff-Pedersen, 1996). Thus, in this model, although the slumped material may have been of glacial origin, seen in rare faceted and striated clasts and dropstones, the diamictites were not directly deposited by ice and so cannot be considered to be tillites. Instead, Arnaud and Eyles (2002) suggested that they record deposition of unstable sediments on the edge of an active extensional basin (but see Edwards, 2004; Arnaud and Eyles, 2004), in which icebergs or sea-ice contributed dropstones. Ice-proximal settings were inferred only for the eastern shore of Skjåholmen, where sediment gravity-flow diamictite is closely associated with well-stratified glaciofluvial sandstones and conglomerates, and at Handelsneset, where complex deformation observed in conglomerate and sandstone sequences suggests active deformation by overriding ice (Arnaud, 2008). From a structural viewpoint, no evidence of such a fault defining a southern basin margin has been documented west of Varangerfjord, although reactivation of an approximately E-W–trending fault would have been extremely likely during the formation of the Barents Sea area in post-Caledonian times.

The proposed source areas, and thus sediment transport (ice mechanism or otherwise) directions for the Smalfjord Formation, also vary. Føyn and Siedlecki (1980) argued that the Middle and possibly also the Lower Krokvatn Diamictite (on Laksefjordvidda; Fig. 3) were derived from the south, based on the presence of crystalline clasts, presumed to be from the autochthonous Baltic Shield to the south. However, as the Kunes Nappe (Fig. 3) also is composed of crystalline basement rocks and restores to a position north or northwest of the Krokvatn paleovalley (Rice, 2001), this argument is invalid. Arnaud and Eyles (2002) also proposed a southerly source area for the probably equivalent rocks at Vieranjar'ga in the Varangerfjord area. Edwards (1984) inferred a westward ice flow along the Varanger paleovalley, with two dominant striation directions documented: NW-SE and E-W (Bjørlykke, 1967; Edwards, 1975; Jensen and Wulff-Pedersen, 1996; Rice and Hofmann, 2000; Laajoki, 2002, 2003).

Edwards (1984) proposed that of the five advance-retreat cycles identified in the Tanafjord area, the first was southerly derived, the subsequent three were northerly derived, and the last is of unknown origin. The general flow or transport directions were based on a comparison of the till composition with that of the immediate subcrop, and the presence or absence of dolomite and/or basement clasts (Edwards, 1972, 1984; note that this also applies to the Mortensnes Formation). Hansen (1992), in contrast, suggested that the lower part of the succession on Gæssenjar'ga had a NE provenance, while the latter part had a SE provenance. These differences are not simply due to the different order of the lithological units proposed by Edwards (1984) and Hansen (1992).

The carbonates at the base of the Nyborg Formation, directly above the Smalfjord Formation, show a mix of dolostones and generally red clastic sediments, forming Member A. More massive dolostones (facies NA1) show sheet cracking and the formation of isopachous cements. These factors, together with the stable isotope data (see following), are strongly suggestive of a Marinoan cap dolostone. The interlayered red shale and dolostone succession is much less typical of such cap-dolostone units, and here sheet cracking is absent in the generally <10-cm-thick dolostone layers, although these show the same stable isotope values as the massive dolostones. In some areas, the massive dolostones gradually change into interbedded sequences along strike (Edwards, 1984). This suggests that variations in current strength and the supply of red clastics compared to dolomite were the controlling factors on the style of cap dolostone that formed in a rugged and variable paleogeography. The red shales are taken to reflect weathering of the steep, very proximal basement topography in this area, still visible to the south of Varangerfjord.

The Mortensnes Formation represents two advances and retreats of ice (Edwards, 1984); the first was initially derived from the south (Lower Member) and then from the north (Middle Member). The source area of the second advance (Upper Member), which shows a major change in provenance, is not known. For both cycles, lodgment tillite was followed by floating ice, giving finer-grained sedimentation and dropstones. The polymict conglomerate draping the formation has been interpreted as a lag-conglomerate, formed during local isostatic uplift, after glacial retreat (Edwards, 1984). The succession passes conformably upward into the postglacial Lillevatn Member, representing fluviodeltaic and subsequent marine deposition.

Stable Isotopes and Neoproterozoic Correlations

Indirect estimates of the age of the rocks have been made by comparing $\delta^{13}C$ data from East Finnmark with the relatively well-documented $\delta^{13}C$ curve for the Neoproterozoic; this is constrained by several precise isotopic ages (cf. Halverson et al., 2005; Halverson and Shields-Zhou, 2011; Condon and Bowring, 2011). Four significant negative $\delta^{13}C$ anomalies are relevant to the Finnmark area: (1) The Bitter Springs anomaly, the onset of which is dated to 810 Ma, reached values close to $-5‰$ $\delta^{13}C$ (Vienna Peedee belemnite [VPDB]); (2) the Trezona anomaly occurred directly before the Marinoan glaciation, with values reaching close to $-10‰$ $\delta^{13}C$ (VPDB); (3) the Marinoan cap-dolostone anomaly (Maieberg anomaly), constrained to 635.5 ± 0.2 Ma (cf. Condon and Bowring, 2011), reached values around $-7‰$ $\delta^{13}C$ (VPDB); and (4) the Wonoka anomaly, which reached values as low as $-12‰$ $\delta^{13}C$ (VPDB), is temporally associated with the 584–582 Ma Gaskiers glaciation (Condon and Bowring, 2011; all $\delta^{13}C$ values from Halverson et al., 2005; Halverson and Shields-Zhou, 2011). Two other anomalies, pre- and postdating the Sturtian glaciation (that pre-dated the Marinoan glaciation), also reached $\delta^{13}C$ values approaching $-5‰$ (VPDB), but neither these, nor the Sturtian glaciation (in whatever form that took), have been found in the Finnmark area.

Dolostones at the base of the Nyborg Formation, directly overlying the Smalfjord Formation, have negative $\delta^{13}C$ values ($-1.0‰$ to $-5.9‰$ VPDB; Halverson et al., 2005; Rice and Halverson, 2005–2010, personal observations), and, where massive, have sheet cracks, pseudo-tepees and, at one locality on Ruos'soai'vi, very small sulfate crystal fans (now replaced by quartz). These characteristics, in combination with their buff-weathering color, strongly suggest that this is a Marinoan-type cap dolostone (cf. Kennedy et al., 1998). U-Pb dating of volcaniclastic deposits in comparable rocks in Namibia gave an age of 635.5 ± 0.2 Ma (cf. Condon and Bowring, 2011). This constrains the age of the basal part of the Nyborg Formation and places a constraint on the youngest possible age of the Smalfjord Formation.

The $\delta^{13}C$ data from the 230-m-thick Grasdal Formation in the Trollfjord area, stratigraphically below the Smalfjord Formation (Figs. 3 and 4), gradually rise from initially negative values (–2.5‰ VPDB) to positive values (6‰ VPDB) before falling rapidly to negative values (–2.5‰ VPDB) and then recovering to positive values (4‰ VPDB) directly below the ~20-m-thick Fadnuvaggi formation, which lies directly and unconformably (Edwards, 1984) under the Smalfjord Formation in the Grasdal area. Halverson et al. (2005) equated this $\delta^{13}C$ pattern with the Trezona anomaly that underlies Marinoan glacial deposits elsewhere (Halverson and Shields-Zhou, 2011) and used this as a further reason to correlate the Smalfjord Formation with the Marinoan glaciation.

A broadly similar anomaly also occurs in the Lossit Formation under the Port Askaig Formation in the Scottish Dalradian rocks (Prave et al., 2009). Based on outcrops in the Irish Dalradian, McCay et al. (2006) argued that the Port Askaig Formation was not equivalent to the Marinoan glaciation, casting doubt on the use of the Trezona anomaly as a criterion for identifying Marinoan deposits. More recently, Rooney et al. (2011) used Re-Os dating to show that the Port Askaig Formation is younger than 659.6 ± 9.6 Ma, tending to support the original Marinoan-age interpretation and hence placing a constraint on the oldest age of the Smalfjord Formation. Despite these uncertainties concerning the significance of the Trezona anomaly, the cap-dolostone characteristics secure the Smalfjord Formation as a much more likely equivalent of the 636.3 ± 4.9 Ma and 635.5 ± 1.2 Ma (Zhang et al., 2008; Hoffmann et al., 2004) Marinoan glacigenic succession. These ages thus constrain the age of the Smalfjord Formation.

Thin dolostones in the upper part of the Nyborg Formation (Member E; Figs. 4, 6, and 7), with $\delta^{13}C$ values of –7.6‰ to –9.9‰, are likely correlatives of the Wonoka anomaly, which elsewhere has $\delta^{13}C$ as low as –12‰ (VPDB) and is the only negative $\delta^{13}C$ value of this dimension known from Ediacaran times (Halverson et al., 2005; Halverson and Shields-Zhou, 2011; Le Guerroué et al., 2006). This places a broad constraint on the upper age of deposition of the Nyborg Formation. Further, it implies that the almost immediately overlying Mortensnes Formation is broadly a correlative of the 584–582 Ma Gaskiers diamictite (Bowring et al., 2003; Halverson et al., 2005; Halverson and Shields-Zhou, 2011).

The $\delta^{13}C$ values from the dolomitic diamictite at the top of the Mortensnes Formation are also strongly negative (–8.4‰ to –10.4‰; Rice et al., 2011), likely reflecting erosion of the dolostones in Member E of the Nyborg Formation.

NATURAL HISTORY AND CULTURAL BACKGROUND

The vegetation in the excursion area varies between occasional wooded-tundra in sheltered low-lying areas and tundra in exposed and upland areas. Clitter, consisting of angular, up to meter-sized loose blocks of the directly underlying lithology, is typical of upland areas, especially where underlain by quartzites/sandstones.

The most common tree is the birch (*Betula* spp.). This can grow to several tens of meters high, although in windy areas, it is completely stunted, with no height, but a very thin "stem" (<1 cm) that may have a lateral extent of several meters. Pine trees (*Pinus sylvestris*) are rare because they have all been cut down, not because the conditions are too extreme. Willow (*Salix* spp.) and less often alder (*Alnus* sp.), poplar (*Populus* sp.), and rowan (*Sorbus aucuparia*) trees are also present. There are several species of *Salix*, but these produce a large array of hybrids, so identifying true "species" is extremely difficult. Low juniper bushes (*Juniperus communis*) occur on well-drained ground, frequently growing over deep holes between boulders; it is, therefore, dangerous to stand on.

On dry gritty slopes, especially below outcrops of the Nyborg and Mortensnes Formations on the north side of Varangerfjord, the unusual ferns *Botrychium lunaria* (Fig. 10A) and rarely *Botrychium boreale* grow. Nyserot (*Veratrum album*; Fig. 10B) is an extremely poisonous lily that is found only in the Tana-West Varangerfjord region and in the Alpine-Apennine-Carpathian region of Central Europe. In contrast, common scurvygrass (*Cochlearia officinalis*; Fig. 10C), which grows in salty areas, is very healthy. As its name implies, it is rich in vitamin C, and was often eaten in earlier times to prevent scurvy, especially by sailors and hunters on Svalbard (e.g., Roberts, 2003). Tana thyme (*Thymus serpyllum* subsp. *tanaënsis*), with small pinkish-blue flowers and leaves having the typical thyme smell, is usually found on dry sandy soils along, for example, the Tana River, while Swedish dogwood (*Cornus suecica*) is an attractive common white flower often forming a significant ground cover. The most common orchids are *Dactylorhiza maculata* spp., quite commonly seen on boggy ground, such as along the Tana River. On ground underlain by carbonate rocks, *Gymnadenia* spp. are the most common orchids (Fig. 10D). The "reindeer lichen" (*Cladonia* spp.; Fig. 10E) is but one of many common lichens eaten by reindeer. Note that picking wildflowers is illegal.

Four edible berries are common. The yellow-orange cloudberries (moltebær; *Rubus chamaemorus*), a Scandinavian delicacy, are locally abundant on peat tussocks in and around boggy ground (Fig. 10F). Bilberries (blueberries, blåbær; *Vaccinium myrtillus*) and especially crowberries (krekling; *Empetrum nigrum*) are common tundra plants; the former taste excellent, while the latter are edible but rather tasteless. Bog (or northern) bilberry (blokkebær; *Vaccinium uliginosum*) look very similar to bilberries, but they are somewhat bigger and have a tasteless and colorless juice; for some people, they can be slightly poisonous. Shiny red berries are cowberries (lingonberries; tyttebær; *Vaccinium vitis-idaea*).

From 2003 to 2008, there was a very serious plague of moths, the caterpillars of which ate all the leaves of most plants; the population grew because winter temperatures failed to reach −30 °C for long enough to kill the larvae. In spring, the moths were so abundant that it appeared to be snowing. Where the population multiplied unchecked, the caterpillars ate and killed off all the vegetation, including the ground vegetation, although poplar and pine trees were unaffected. These damaged areas now appear rather blackened and will take ~30 yr or more to recover fully.

Sea-fishing (especially cod; *Gadus* sp.) was a major industry up until the mid-1970s. However, a rapid and catastrophic decline in fish stocks in the mid- to late 1970s led to an industry collapse, and wooden boats were abandoned on beaches to rot. As a result of this decline, fish factories were no longer used and have mostly collapsed; previously, the quays at Vadsø, Vardø, and

Figure 10. Finnmark plants. (A) *Botrychium lunaria*, an unusual fern; 5–25 cm. June-August. (B) Nyserot *(Veratrum album)*, an extremely poisonous lily; 60–120 cm. June–August. (C) Common scurvygrass *(Cochlearia officinalis)*, a rich source of vitamin C; 10–35 cm. May–June. (D) *Gymnadenia conopsea*, a common orchid on carbonate ground; 10–40 cm. June–August. (E) *Cladonia* spp., a form of "reindeer lichen"; up to ~20 cm across. (F) *Rubus chamaemorus*, the cloudberry; 10–25 cm. June–August.

Hamningberg (among many other places; Fig. 1) were crowded with fish-processing factories, built out over the water so as to be accessible at low tide. Even small villages had a local industry. The business has partly recovered in the past 15 yr. The decline was due both to overfishing and to the fish moving to other areas, as is known to happen. In the old days, many fish caught in the winter were hung up on wooden racks, often tent shaped, to dry by sublimation in the freezing air; these structures are still occasionally seen along the shoreline, as at Nesseby (Fig. 1). Such *stockfish* are still produced, but in lesser amounts. They are eaten either after boiling or, more commonly, after being hammered into small pieces.

King crabs (*Paralithodes* sp.), which are not true crabs but are closely related to hermit crabs (also not true crabs), were originally brought from Kamchatka to the Kola area by the Soviets. They are now an important part of the fishing industry. Although previously they were protected from overfishing, they are no longer safeguarded since they damage the environment for local species.

The Tana River (Fig. 1) is an important European salmon river, with between 70 and 250 tonnes (metric tons) (average of 140) of fish caught each year. If the salmon caught in Tanafjord is included, the total increases to 600 tonnes, making the Tana system the largest source of wild North Atlantic salmon (*Salmo salar*). However, a report in 2010 noted that stocks have dropped alarmingly in the last few years, and discussions on new regulations started in 2011. The Tana is unusual in that people who live along the riverbank have the traditional right to catch salmon in nets, from late May to the end of August. Elsewhere, only rod fishing is allowed in rivers, although fixed coastal net fishing is used in fjords. Both river and marine nets have to be checked regularly (twice or more daily), to prevent damage to the fish caught, as this destroys their commercial value, and can only be set for four days a week, to allow some fish to reach the breeding grounds. In the late 1980s and 1990s, almost every fjord appeared to have one or more salmon fish farm. Most of these soon closed down due to overproduction and disease.

Reindeer (*Rangifer tarandus*) are commonly seen in the summer as small herds of females with calves, or solitary males. Herds get bigger in the autumn. All reindeer belong to Saami people; the owner can be identified by the shape of the ear-clipping. Elk (*Alces alces*) are also surprisingly numerous but are rarely seen. Sheep are common in grassy coastal areas, often sleeping on the dry ground at the roadside; they should be treated with great caution when driving, as they tend to walk directly in front of fast-moving cars. There are also very rare wolverines (*Gulo gulo*) and bears (*Ursus arctos*), as well as red foxes (*Vulpes vulpes*) and arctic foxes (*Alopex lagopus*). The latter have a gray-white tail during the summer, though they are otherwise typically fox-red in color. Lemmings (*Lemnus lemnus*) are extremely abundant in some years; 2007 and 2011 were both "lemming years." Ekkerøya (Fig. 1) hosts a major bird colony (seagulls and kittiwakes; *Rissa* sp.), and many coastal areas are home to arctic (*Sterna paradisaea*) and common terns (*Sterna hirundo*). The sea-eagle (*Haliaeetus albicilla*) is making a major comeback through active conservation, after nearly becoming extinct toward the end of the twentieth century.

There is evidence of stone-aged human inhabitation in the area since 10,000 yr B.P., with settlements moving downhill to keep pace with the postglacial isostatically retreating shoreline. More details of this can be obtained at the Varangerbotn and, especially, the Mortensnes museums (Fig. 1); Mortensnes is the site of superb Neolithic graves. The Saami, or Sámi, people have lived in northern Scandinavia-Russia (Sápmi) for at least 2000 yr, with peoples from the south arriving more recently. Tacitus and Ptolemy referred to *Fenni* and *Phinnoi* peoples, respectively, who might have been the Saami, but this cannot be unequivocally substantiated or refuted. There are currently 80,000–135,000 Saami, well over half living in Norway; the rest live in Russia, Finland, and Sweden. The Saami parliament has authority to deal with some transborder Saami matters. Each Saami group has its own variation of the local dress, seen in the shape of the hats or in the design of the highly decorated borders of dresses and jackets, the *gákti*. The ten Saami languages used in the Sápmi area, which are not mutually intelligible, are all part of the Finno-Ugric branch of Uralic languages.

The Saami flag (Fig. 11) was inspired by a shaman's drum and the poem "Päiven parneh" written by Anders Fjellner (1795–1876), which described the Saami as the children of the Sun through the union of a female giant, who lives in a "House of Death" far in the north, and the Sun's male offspring, with whom she elopes. The flag's circle represents the sun (red) and the moon (blue), with the traditional Saami colors, red, green, yellow, and blue.

Originally, the Saami lived in the interior, and the relative newcomers from the south lived along the coast. However, the Black Death (1349) reduced the coastal population so much that the Saami were offered incentives to move to the coast and become fishers/farmers. This continued up to the early 1700s. Consequently, only ~10% of the Saami are now "Mountain Saami," involved in reindeer herding; most are "Sea-Saami," living in coastal areas.

Most buildings in East Finnmark were burnt down during the retreat of the German army in 1944. Buildings that were not destroyed can be seen in Hamningberg, in and near Vadsø,

Figure 11. The Saami flag.

and at Kongsfjord (Fig. 1). They can be recognized by being more ornate, for example, around the windows and on the eaves and gables. Houses built in the period between the war and the "recent" oil-revenue boom tend to be cube-shaped and sparse. Boom-time houses have more of an "alpine-chalet"–type architecture. Houses are invariably made of wood, with a concrete cellar, and are superbly insulated.

The wooden church at Nesseby (Fig. 1), which escaped destruction at the end of the war, is a noted photogenic feature, lying on a peninsula on the north coast of Varangerfjord. It was built in 1858 (the original church dated to 1747) and has a chapel dating to 1719 beside it. Cast-iron crosses in the graveyard date back to the 1780s, when the trade between NW Russia and northern Norway, part of the Russian-based Pomor trade, was relatively new, having started in 1740. Similar crosses are present in the graveyard at Polmak, near the Finland-Norwegian border, the church of which dates to 1853, with an altar piece from 1626. The Pomor trade survived until the Russian revolution of 1917. Relicts of Pomor buildings lie along the Varanger coast; those at Handelsneset, which means Trading-point, near Mortensnes, were owned by the Nordvi family. They were part of a trading post that was founded in 1784 and went bankrupt in 1877. The houses were then sold and removed, leaving only the foundations.

The expansion of the road network has had a major effect on the distribution of inhabitants. Prior to the 1970s, there were many small communities around the coast (Fig. 1), serviced by regular small local ferries. With the arrival of roads, and likely hastened by the decline in the fishing industry, many communities were abandoned, or are still in the process of being abandoned, with people moving to places on the road network with more inhabitants or better infrastructures. This process was more extreme in the Tanafjord area than around Varangerfjord (Fig. 1). In the abandoned settlements, some of the dwellings have been maintained for summer or holiday use, and, in recent years, new summer huts (often very luxurious) have been constructed.

There are museums (Fig. 1) at Polmak (Tana fishing history and local culture), Varangerbotn (Saami lifestyle), Mortensnes (a part of the Varangerbotn Museum; prehistoric graves), Vadsø (Norwegian daily life, farming, and lifestyles), Vardø (1730s (fortifications), Byluft (local lifestyle and fishing), and Kirkenes (mining, World War II, art); all are well worth visiting. Further afield, there are also museums at Båtsfjord (local history), Berlevåg (World War II, fishing and local history), Karasjok (Saami culture, including a collection of traditional/old buildings), and, a day's drive away, Alta (United Nations Educational, Scientific, and Cultural Organisation [UNESCO] World Heritage Site; spectacular Neolithic rock carvings; Fig. 1). These, too, are all worth visiting.

EXCURSION ROUTE AND ROAD LOG

The order of the days given is arbitrary, since weather-dependent boat journeys are included; these should be done at the earliest possible moment.

The use of hammers anywhere is strongly discouraged unless samples are being taken as part of a proper scientific study or from an outcrop where the material taken is present in great abundance. At Oaibaččannjar'ga, the use of hammers is entirely prohibited, by law. Further, a license is required to export rocks, minerals, and fossils from Norway.

Global positioning system (GPS) coordinates are given for most outcrops. "GEC" refers to coordinates derived from Google Earth. The two systems give the same coordinates, but the latter are less accurately positioned, largely because Google Earth currently has poor quality satellite images of East Finnmark. However, to make finding the outcrops easier, the positions of all the stops and locations can be plotted on Google Earth and the image then downloaded prior to undertaking the excursion. Much better quality images can be screen-dumped from www.1881.no/kart/# or from http://kart.statkart.no.

Day 1. Lower Smalfjord Formation in South Varangerfjord and West Karlebotn, Varanger Paleovalley (I)

Introduction

The southwestern and western parts of Varangerfjord (Figs. 3 and 12), this mostly being the Karlebotn area, have many outcrops showing the unconformable contact between the Smalfjord Formation and the crystalline gneisses of the Baltic Shield. Several such outcrops are included in the excursion on this day. Together with outcrops in the Oaibaččanjar'ga and Handelsneset areas (Fig. 3), where the Smalfjord Formation overlies the Vadsø Group, these define the geometry of the Varanger paleovalley. Often, diamictite is absent or only locally exposed at the smaller outcrops; instead, sandstones are seen to onlap or to be draped over the irregular basement surface, with variable primary dips and dip directions. Evidence for glacial onlap can be seen at the outcrops on Skjåholmen and Handelsneset; this is schematically shown in Figure 6. Glacial striations are developed at some of these outcrops.

The sediments were deposited on a very irregular crystalline basement topography, at a variety of scales. The role of the Pleistocene glaciation in forming the basement topography at the outcrops seen is thought to be relatively small (cf. Bjørlykke, 1967). Where striations or any other sharply irregular basement-cover topography (at whatever scale) is preserved, the Pleistocene contribution was minimal. Equally, little Pleistocene erosion of the basement has occurred where a smooth basement surface is developed between two patches of cover sediments, and this surface is continuous with the basement-cover contact under the sediments (e.g., at Stop 1.4).

The Lattanjar'ga unit sandstones, the most easterly exposures of Neoproterozoic sediments in the south Varangerfjord area are also included. These might be a part of the Smalfjord Formation, in contrast to the older age previously published by Rice et al. (2001), who proposed an age equivalent to or below the Veinesbotn Formation at the base of the Vadsø Group. Since the Veinesbotn Formation is now thought to be part of the Tanafjord Group (Røe, 2003), correlating the Lattanar'ga unit with the Veinesbotn Formation is less certain. Visiting Lattanjar'ga requires a walk of several kilometers.

Stop 1.1: Lattanjar'ga Peninsula

Location. If coming from Kirkenes, drive through Gandvik (Fig. 12), up the steep hill to the west, past the small lake on the north side, and, ~950 m further on, take the dirt road north, at GEC 70°02'31"N, 29°06'37"E. If coming from Varangerbotn, drive through Nyelv and then a further ~10 km. The dirt road is ~850 m past the east end of the big lake immediately south of the road. Drive as far down as you can and park so that other vehicles can pass by with as little damage to the environment as possible, and then walk to the bottom of the hill and continue north and east along the coast for ~750 m, to where the beach abuts rugged basement outcrops (Mak'kagåp'pi; GEC 70°03'02"N, 29°06'15"E). From here, climb up onto the basement and walk north for a further 750 m at ~60 m above sea level (a.s.l.), to the large outcrop of bedded sandstones overlying the basement at Urračåk'ka. This outcrop can be seen clearly from the main road (especially with binoculars) as a small smoothly sloping area overlying the very irregular basement surface.

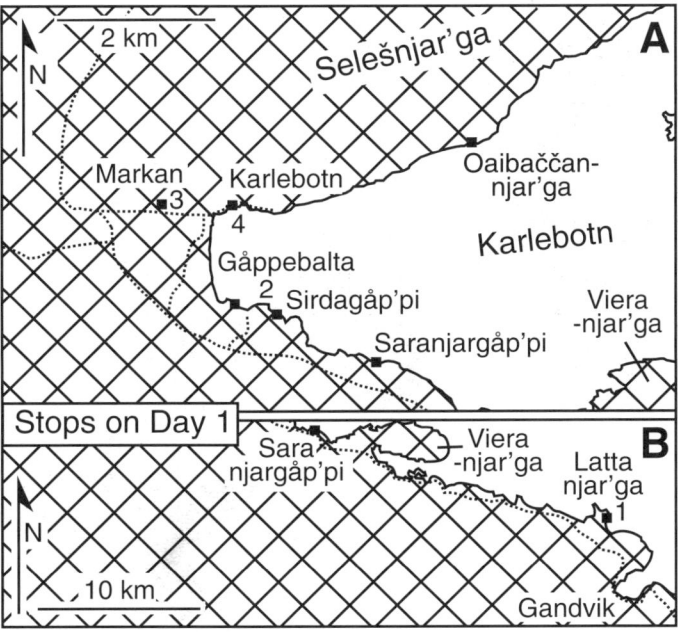

Figure 12. Map showing the location of the outcrops visited on Day 1. Dashed line indicates roads.

Grid ref. NT 8000 7370, 1:50,000 map sheet Nesseby 2335 II, Edition 2-NOR. GEC 70°03'18"N, 29°05'60"E.

Introduction. This is the most easterly recorded outcrop of Neoproterozoic sediments on the south side of Varangerfjord. Their age is uncertain; Rice et al. (2001) placed them within the Vadsø Group, at the base of or below the Veinesbotn Formation, but alternatively they could be part of the Smalfjord Formation, comparable to the sandstones at Sirdagåp'pi (Stop 1.2). If the latter is assumed, then they form part of the major unconformity created within the Neoproterozoic succession by the westerly movement of a Marinoan glacier in the area now occupied by Varangerfjord, forming the Varanger paleovalley (Bjørlykke, 1967).

Description. Although the sandstones described can be studied at Mak'kagåp'pi, it is well worth walking the extra distance north to the largest outcrop. Here, a crudely elliptical outcrop, with up to 12 m thickness of pale pinkish-yellowish to honey-colored quartz arenites in beds 0.2–1 m thick, fills a northeastward-plunging paleodepression in basement gneisses. Overall, the beds form a gentle syncline, also plunging northeast from 60 to 30 m a.s.l. Channeling up to 1 m deep occurs within the sandstones (Fig. 13A), as does low-angle cross-bedding (sometimes herringbone; Fig. 13B) and large (up to 75 cm high) foresets. Laminations in the cross-beds are in some cases picked out by pink layers, likely feldspar concentrations. Symmetrical ripples also occur. Around the whole outcrop, the basal parts of the deposits contain occasional rounded boulders of coarse-grained mafic schists up to a meter in size. The sandstones both onlap and are draped over these boulders (Figs. 13B and 13C), which may overlie up to 1.5 m of sandstone. Rarer, much smaller (<10 cm) rounded quartz pebbles also occur in this basal part. No equivalent of the mafic gneisses has been found in situ in the nearby basement.

At Mak'kagåp'pi, similar sandstones are best exposed at low tide, although overgrown outcrops lie S-SW of the beach, near a prominent scarp in the basement. About 4 m of sandstones are exposed, dipping gently NNE. To the west, the bedding has a variable strike, with dips of up to 65°, sharply juxtaposed by faulting against gently dipping beds along a junction oriented 017/65°E, subparallel to the basement fault, and marked by a few

Figure 13. Lattanjar'ga, south Varangerfjord. All at Stop 1.1. (A) Channel structure in sandstones of the Lattanjar'ga unit. (B) Herringbone cross-bedding in sandstones of the Lattanjar'ga unit, above large mafic gneiss clasts. (C) Sandstones of the Lattanjar'ga unit both onlapping and draped over a large clast of mafic gneiss, near the base of the Lattanjar'ga unit. (D) Red-green shales of uncertain stratigraphic affinity.

millimeters of greenish siltstone. A few meters south, the rocks are folded but without any fault offset.

In the southwest part of the beach, massive greenish-gray sandstones, weathering pale gray and in beds up to 0.75 m thick, are exposed only 1.5 m from the basement. The contact with the yellow sandstones cannot be seen due to the shifting loose beach sand. These gray sandstones have a sharp contact, perhaps faulted, against a small outcrop (0.75 m^2) of red/maroon sandstones (Fig. 13D), lying closer to the basement, with common detrital muscovite and 0.5-cm-sized green reduction spots. This red lithology merges into a pale-green clast-supported conglomerate/breccia composed of angular to rounded fragments (3–4 mm size) of quartz, feldspar, and detrital micas. These gray/red/green sandstones are very different to other parts of the Lattanjar'ga unit, and it is not clear if they are a part of it.

The steep primary dip of the Lattanjar'ga unit, subparallel to the basement topographic surface, suggests a correlation with the Smalfjord Formation; such an orientation is typical of the Smalfjord Formation in this area (see Stops 1.2, 1.3, and 1.4) and atypical of the Veinesbotn Formation. If the correlation with the Veinesbotn Formation is retained and this is placed in the Tanafjord Group, as proposed by Røe (2003), it would indicate that the slight angular unconformity at base of the Tanafjord Group on the north side of Varangerfjord cuts down suddenly through the lower half of the Tanafjord Group, as well as the Ekkerøya and Vadsø Groups, across Varangerfjord. This would likely imply a fault in Varangerfjord, as proposed by Røe (2003).

Stop 1.2: Gåppebalta and Sirdagåp'pi, South Karlebotn

Location. Drive back to the main road and turn west. Just before reaching the western end of Karlebotn, take the dirt road down to the house at Grid Ref 6039 7900 (GEC 70°06′35″N, 28°35′23″E). After asking for permission, walk essentially northward from the house, down to a small bay.

Grid ref. NT 6020 7913, 1:50,000 map sheet Varangerbotn 2335 III, Edition 3-NOR. GEC 70°06′40″N, 28°35′17″E.

Introduction. The south shore of Karlebotn is marked by a number of small outcrops of presumed Smalfjord Formation resting directly on the basement; these dip northward, overall, defining the southern margin of the Varanger paleovalley.

Please do not use hammers anywhere at this outcrop; there is very little material exposed.

Description. On the west side of the bay, a vertical basement surface has up to 0.5 m of overlying cover sediment preserved (Fig. 14A). High up (2–2.5 m) on the south side, partly above the sloping vegetation, the rock surface has been smoothed by the Pleistocene glaciation. Most of this smoothed surface is basement, but elongate "lines" of Smalfjord Formation sandstones are also preserved. These are thought to be relicts of Smalfjord Formation infilling grooves (?glacial striations) in a Neoproterozoic topography (Fig. 15). Members of the excursion in 2008 noted that these grooves are parallel to the gneissic layering in the basement and thus could have been caused by differential weathering prior to Neoproterozoic sedimentation, and thus might not be glacial striations. Grooves oblique to the layering (and thus glacial striations; Fig. 14B) were found by Paul Hoffman at the base of the outcrop, near to the right side (north) of the boulders at the base of Figure 14A. Similar striations were reported at Saranjargåp'pi, a few kilometers to the SE (Fig. 12, GEC 70°06′16″N, 28°37′55″E; Rice and Hofmann, 2000).

On the same basement surface, down at the level of the beach sand, the immature sandstone cover sediments show a vertical compositional banding (?bedding; Fig. 14C). This orientation may be primary, perhaps caused by smearing of sediment against the vertical basement surface as ice moved past this steep surface. Alternatively, it might be secondary, formed by the initial deposition of relatively steeply dipping sediments against the steep basement paleotopography, with the dip angle accentuated by subsequent compaction and dewatering during lithification; similar deposits have been seen elsewhere, although not with such steep orientations (see next three paragraphs and also Stop 2.3). Finally, but most unlikely, the steep orientation might be of tectonic origin, indicating that the basement here is allochthonous (as it is elsewhere; see Introduction to Stop 2.3).

On the rounded, north-facing basement surface, ~50 m west of the steep face, small patches of cover sediments are molded to the irregular basement topography over some 4 m vertically, down to sea level, and dipping to the north; "compaction synclines" are also developed here (cf. Fig. 16C from Stop 2.3). Locally, very thin (<1 cm) and small scabs of breccia (?tectonic) lie on the basement surface; these may relate to post-Caledonian deformation.

Some of the sediments lying on the basement carry millimeter-scale crenulations; these are typical of such sediments in the Karlebotn area. Crenulation axes are usually subparallel to the local strike of the sediment in which they occur, and thus the local strike orientation of the basement surface. In thin section, thin pelitic layers (including detrital muscovites) can be seen to be wrapped around detrital quartz and feldspar grains; in two dimensions, this could be a compaction feature, but this would not give linear (three-dimensional) crenulation axes. More likely, the microfolds reflect minor soft-sediment downslope slumping/shortening.

At Sirdagåp'pi, some 450 m to the east (Fig. 12, Grid Ref NT 6070 7895, GEC 70°06′37″N, 28°35′49″E), a considerably larger outcrop (20 × 60 m) of white to pale-yellow sandstone in beds up to 1.75 m thick and with occasional 10 cm basement-derived lonestones has been preserved. The upper sandstones have a more yellowish color, comparable to that of the Lattanjar'ga unit at Stop 1.1. This succession overlies the basement (amphibolitic to west, granitoid to east) on an extremely irregular surface; the bedding is molded to the paleorelief, with dips of up to 60° (Fig. 14D). A 20-cm-thick, fissile, fine-grained matrix-supported basal conglomerate with spherical to subangular quartz clasts up to 0.5 cm size and thin, greeny-blue shale clasts is locally developed (Fig. 14E).

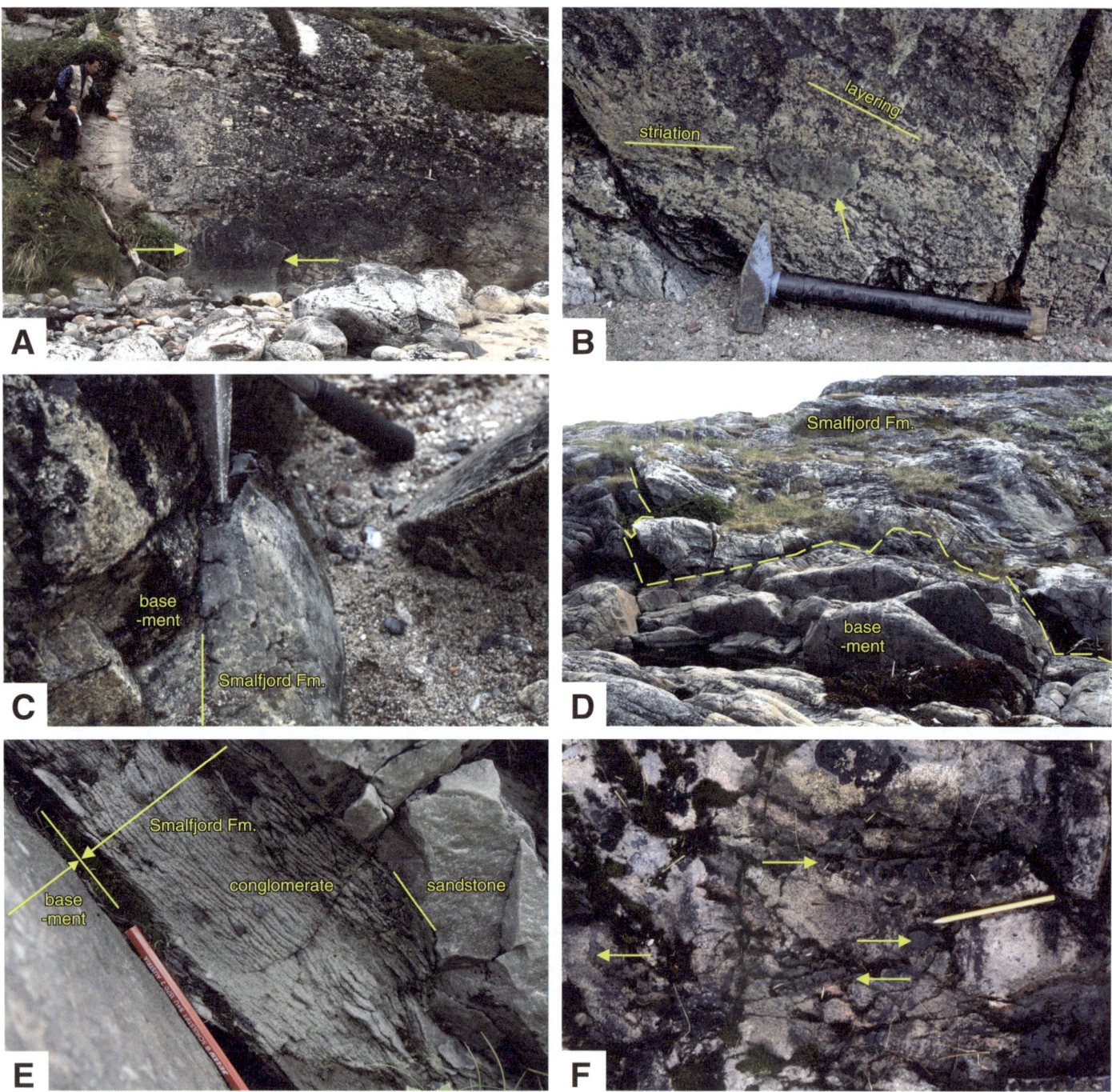

Figure 14. East Karlebotn. All photos are facies S3 of the Smalfjord Formation overlying crystalline basement. (A) Stop 1.2. Man pointing to where grooves occur within basement at Gåppebalta. Note the slab of cover sediment (arrowed). (B) Stop 1.2. Close-up of glacial striations at Gåppebalta, oblique to gneissic layering and filled with cover sediments (arrowed); from the base of and a bit to the right (north) of photo in A. (C) Subvertical layering (bedding) in cover sandstones at Gåppebalta. (D) Stop 1.2. White cover sediments draped over pink Archean basement rocks at Sirdagåp'pi, east of Gåppebalta. Dashed line gives broad outline of the very irregular basement-cover unconformity. (E) Stop 1.2. Thin-bedded, fine-grained conglomerate at the base of the white sandstones at Sirdagåp'pi. (F) Stop 1.4. *Very* small scabs of dark-gray cover sandstones (arrowed) lying on pinkish basement rocks on the north side of the road at Karlebotn.

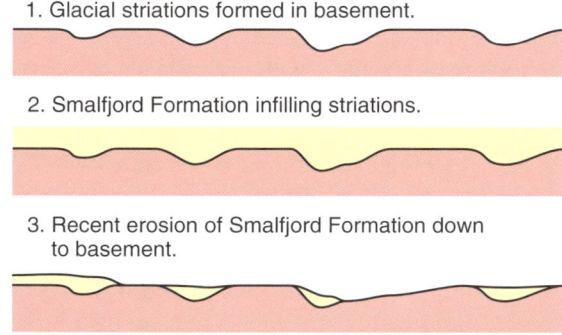

Figure 15. Stop 1.2. Schematic model for the preservation of glacial striations at Gåppebalta.

Stop 1.3: Mar'kan (Karlebotn Village)

Location. Return to the main road and turn west. Drive for ~3 km and then take the turning to the right, down the road trending E-W, toward Karlebotn village. Park after ~0.75 km, essentially at the east GEC coordinate given below (see Fig. 12). After asking for permission in the nearby house, walk north across the fields, for ~70 m, to the relatively distinct rounded outcrops, formed of basement gneisses (east end) and Smalfjord Formation sediments (west end).

Grid ref. NT 5935 8045, 1:50,000 map sheet Varangerbotn 2335 III, Edition 3-NOR. GEC 70°07′25″N, 28°33′22″E.

Introduction. This is one of several places in the Selešnjar'ga area where circular to elongate "dome"-shaped outcrops of crystalline Archean basement rocks are exposed through sandstones of the Smalfjord Formation (Siedlecka, 1990; Fig. 3), well above sea level, which is the lowest level at which the basement-cover contact can be observed (see Stops 1.1, 1.2, 1.4). Other domes exposing the basement-cover contact are documented at Stop 2.3. The smooth nature of the basement surface under the cover sediments suggests glacial smoothing prior to sandstone deposition; no evidence of Neoproterozoic weathering of the basement (a regolith) under the sandstones has been observed.

Description. The basement outcrop is several tens of meters long E-W by several meters wide N-S, with an obvious rounded outcrop in the middle. Around the eastern and southeastern parts of the outcrop, and less often on the northeastern side, the contact with the Smalfjord Formation is exposed. Overall, the basement has a convex-upward surface, but locally it can be seen that Smalfjord Formation sediments infill depressions in the basement. Some depressions are smooth (Fig. 16A), while in others it can be seen that blocks of basement were plucked or fell out and the resulting angular surface irregularity was infilled with sediments. Infilling sediments, which are up to 25 cm thick, include white-weathering sandstones, locally with a sedimentary breccia, comprising a green-blue/gray sandy matrix with angular to

Figure 16. East Karlebotn. All facies S3 of the Smalfjord Formation overlying crystalline basement. (A) Stop 1.3. Smooth depression within the basement at Mar'kan. The yellow arrows point to the contact between the cover sediments, which lie in the smooth depression, and the basement. More sediment is visible above the hammer head and in the channel to the right, under the lichens. (B) Stop 1.4. Thin and small scabs of cover sediment on the basement at Karlebotn (solid arrows). Open arrows point to areas where sediments were until very recently. (C) Stop 2.3. "Compaction syncline" in cover sediments at the basement outcrop NW of Larajœg'gi.

subrounded basement and sandstone clasts up to 10 cm size. On an outcrop scale, both here and at the other nearby basement inliers, the bedding in these infillings, and in the rocks more generally surrounding the basement, dips away from the basement in a "radial" distribution.

Stop 1.4: Karlebotn School

Location. Continue driving downhill from the last outcrop, past the large school building, and park by the sea. Walk northeast on the dirt road for a few meters. The outcrop is an obvious smooth exposure of basement rocks lying just northeast of the parking lot, between the dirt road and the sea and, on the north side of the road, forming a small "bluff" of basement rocks.

Grid ref. NT 6005 8037, 1:50,000 map sheet Varangerbotn 2335 III, Edition 3-NOR. GPS 70°07′23.8″N, 28°34′58.8″E.

Introduction. Please do not use hammers anywhere at this outcrop; there is very little material preserved. This is the most easily accessible outcrop showing the three-dimensional (3-D) nature of the contact between the basement and the cover. To see all the material described, it should be visited at low tide. Here, the continuity of the basement-cover surface under relict patches of cover indicates that Pleistocene glacial erosion did little more than scrape the Neoproterozoic cover sediments off the underlying basement gneisses.

Description. The smooth outcrop consists of pinkish gneisses of Archean age. These are cut by and folded with several generations of pegmatite veins. On the northeast side, the outcrop rather suddenly slopes down to the intertidal mud. On this slope, several patches of bluish gray sandstones and grits, only a few centimeters in size and less than 1 cm thick, have been preserved (Fig. 16B). In Figure 16B, the three patches of cover sediment on the upper-left side reflect the overall basement-cover surface, also preserved in the areas between. Similarly, the single patch on the upper right reflects the overall basement-cover surface, which has a yellow-orange color. The intervening pale-gray area with a rougher surface likely represents an area affected by subsequent erosion (possibly by plucking of large blocks, since the surface is uneven). Below the patch on the right (north side) in Figure 16B, there are several areas where it can be seen that cover sediments were preserved until recently (open arrow).

On the area below the slope, numerous small patches of sandstone all preserve the same smooth basement-cover orientation; no significant Pleistocene erosion of the basement occurred here. Note that the two Neoproterozoic basement surfaces are linked by a sharp change in orientation.

By analogy with nearby outcrops, the sandstones are taken to be a part of the Smalfjord Formation. Small (centimeter-size) outcrops of this sandstone can also be found between the seaweed, at low tide, within the adjacent intertidal muds on the immediately NE side of the outcrop (essentially below the hammer in Fig. 16B).

At low tide, seaweed- and mud-covered basement continues for several meters to the southwest from the main outcrop, with small patches of the Smalfjord Formation. Also, some 100 m to the east, a 30-m-long E-W–oriented outcrop exposed at very low tide also comprises basement with occasional small silty-sandy patches of the Smalfjord Formation hidden under the weeds.

On the north side of the road, slightly to the west, there is an ~10-m-wide by 5-m-high outcrop of basement rocks. The steep surface of this outcrop is a primary basement-cover contact; in the middle, near the base, small (centimeter-size) relicts of the Smalfjord Formation are preserved on the basement, dipping (with the eye of faith!) steeply southward (Fig. 14F).

When combined with the basement gneiss platform visited, these outcrops show the intermediate-scale, uneven nature of the basement-cover surface.

Day 2. Lower Smalfjord Formation at Oaibaččanjar'ga and Selešnjar'ga, Varanger Paleovalley (II)

Introduction

During this day, an outcrop of the lower part of the Nyborg Formation is examined, as well as exposures at or near the base of the Varanger paleovalley on the northern side of Karlebotn, at the western end of Varangerfjord (Fig. 17). In one case, this will be the contact between the Precambrian basement and the Smalfjord Formation, as seen the previous day. In another case, at the famous outcrop at Oaibaččannjar'ga (Bigganjar'ga), the base of the paleovalley lies within the Veinesbotn Formation, with spectacularly developed glacial striations; this stop requires considerable time for a proper investigation and is best done at low tide.

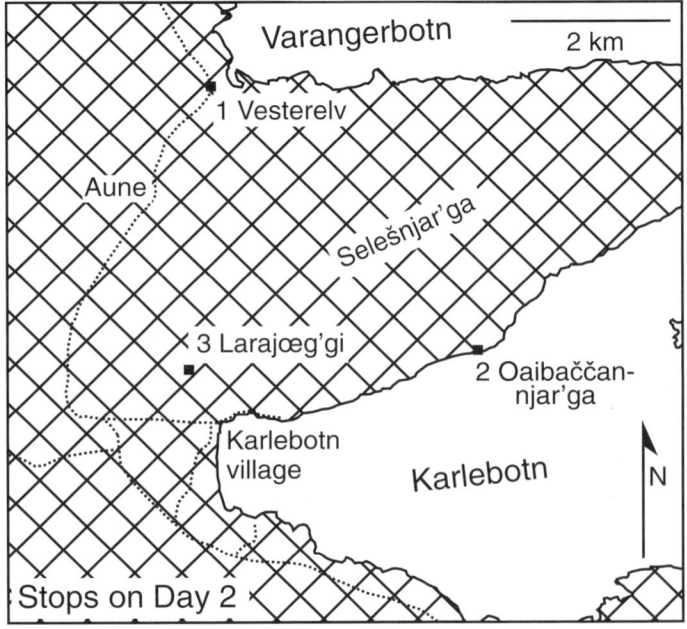

Figure 17. Map showing the location of outcrops visited during Day 2. Dashed line indicates roads.

Stop 2.1: Vesterelv

Location. After turning south at the Varangerbotn roundabout, toward Kirkenes, drive for ~1.5 km. Park beside the road where safe and walk to the obvious folds on a sharp right-hand bend just north of Vesterelv.

Grid ref. NT 5955 8450, 1:50,000 map sheet Varangerbotn 2335 III, Edition 3-NOR. GEC 70°09′35″N, 28°34′20″E.

Introduction. This outcrop shows the geometry and orientation of the pervasive tectonic deformation seen within the Nyborg Formation (Member B; Fig. 7). Although likely of Caledonian age, this E-W fold orientation is markedly oblique to the regional N-S to NE-SW trends (Williams, 1979). Such folds have been recorded at almost all outcrops of the Nyborg Formation in the Varangerfjord coastal area, including those on the north coast of western Selešnjar'ga (Fig. 3). The amount of shortening accommodated by the E-W folding in the Nyborg Formation is considerable; such folds outcrop from the Tana Bru area to south of Vesterelv, within the boggy ground, right up to the hills made up of Archean rocks of the Baltic Shield. Although the bedding gets much more massive to the south, and fold wavelengths increase, the total amount of shortening within the Nyborg Formation is in the order of 60% (Townsend, 1986), giving a possible maximum, essentially southward-directed, shortening of up to 11 km. This needs to be kept in mind when at Stop 2.3.

Description. Red/purple mudstones are interbedded with fairly thick-bedded gray sandstones, typical of Member B of the Nyborg Formation. The latter sediments are in parallel-sided, poorly sorted, and massive to fining-upward graded beds. This has been interpreted as a relatively deep-water deposit of a submarine fan (Edwards, 1984). The rocks have been deformed by flexural slip into south-vergent, upward-facing, close to tight folds with moderately dipping axial surfaces and long, planar limbs and narrow hinge zones (chevron folds). Fold axes trend approximately E-W.

The massive sandstones have a fracture cleavage subnormal to the layering, within which quartz was deposited, while a cleavage subparallel to the axial surface formed in pelitic layers, sometimes with small crenulations in the hinge zone. Mudstones are thickened in the hinges and accommodation structures have developed. Note that these rocks, and their continuation into the boggy ground to the south, lie physically below the stratigraphically older rocks of the Smalfjord Formation exposed on Selešnjar'ga to the SE (Fig. 3). Whether this is a paleogeographic phenomenon (filling the Varanger paleovalley in post-Smalfjord Formation times) or a consequence of Caledonian deformation is not known.

At the south end of the outcrop, a moderately to gently northward-dipping thrust fault has cut through a fold. Slickensides on the thrust indicate movement toward the east, markedly oblique to the fold axes and suggesting that the essentially E-W shortening seen in the bulk of the Gaissa Thrust Belt (Townsend et al., 1986) postdates the N-S shortening seen in the Varangerfjord area.

Stop 2.2: Oaibaččannjar'ga (Bigganjar'ga, Cape Headache)

Location. Drive south from Vesterelv and, after going uphill to the left (east) past massive white sandstones, turn left to Karlebotn and go straight downhill to the sea (Fig. 17). Park just beyond the large building, the old school (GEC 70°07′23″N, 28°34′48″E). From here, walk NE beside or near the sea for ~3.7 km, initially along the dirt road, until you reach the obvious outcrop of diamictite overlying a striated pavement, just before the small lighthouse. The walk takes ~45 min. This outcrop should be visited at low tide, so, depending on the tide level, it may be better to do Stop 2.3 before this stop.

The use of hammers anywhere at this locality/outcrop is prohibited by law.

Grid ref. NT 6320 8133, 1:50,000 map sheet Varangerbotn 2335 III, Edition 3-NOR. GPS 70°07′50.5″N, 28°39′43.4″E.

Introduction. The description of this outcrop by Reusch (1891) is one of the earliest publications to recognize and describe in detail features formed during a pre-Pleistocene glaciation (Hofmann, 2011). Reusch (1891) documented a diamictite overlying a planar platform of quartzites covered with conspicuous striations of two main orientations.

The outcrop was interpreted by Reusch (1891) as a tillite overlying a glacially striated surface. Soon after, other workers visited the outcrop, which was an extremely arduous undertaking in those days, and published reports (Strahan, 1897; Schiøtz, 1898; Dal, 1900). Strahan (1897) stated that Reusch's conclusions had been rejected by the geological community in the face of convincing evidence (Schiøtz, 1898; Dal, 1900). In subsequent years, the outcrop was described regularly (Holtedahl, 1918; Rosendahl, 1931, 1945; Føyn, 1937; von Gaertner, 1943; Spjeldnaes, 1964; following references).

The outcrop rapidly became well known; Rosendahl (1945, p. 336–337) wrote about the geologist Sederholm that, "Of all the geological occurrences he (Sederholm) had seen, this was the most remarkable and revealing. For him, it was a spiritual place, and the journey was a pilgrimage: he used this word himself. This was now the seventh time he was making the journey" (translated from Norwegian by Marc Edwards).

Bjørlykke (1967) interpreted the outcrop as the base of a glacially formed paleovalley, the Varanger paleovalley, along the present-day Varangerfjord. Subsequently, Edwards (1975) published the first detailed sedimentary description of the lithologies.

Description. On approaching the outcrop from the SW, it is immediately apparent that there are three parts to be considered; the basal sandstones/quartzites with striations, a lensoid diamictite, and the overlying onlapping and draping sedimentary sequence, passing up into the bulk of the Smalfjord Formation in the Varangerfjord area (Fig. 18). The slight lithological differences between the basal and overlying sandstone units have been used to trace the boundary for several kilometers southwest of Oaibaččannjar'ga (Rice and Hofmann, 2000; Laajoki, 2002).

The sandstones and quartzites below the diamictite are slightly to relatively strongly pink or gray colored, locally with well-developed tabular cross-bedding. These sandstones were

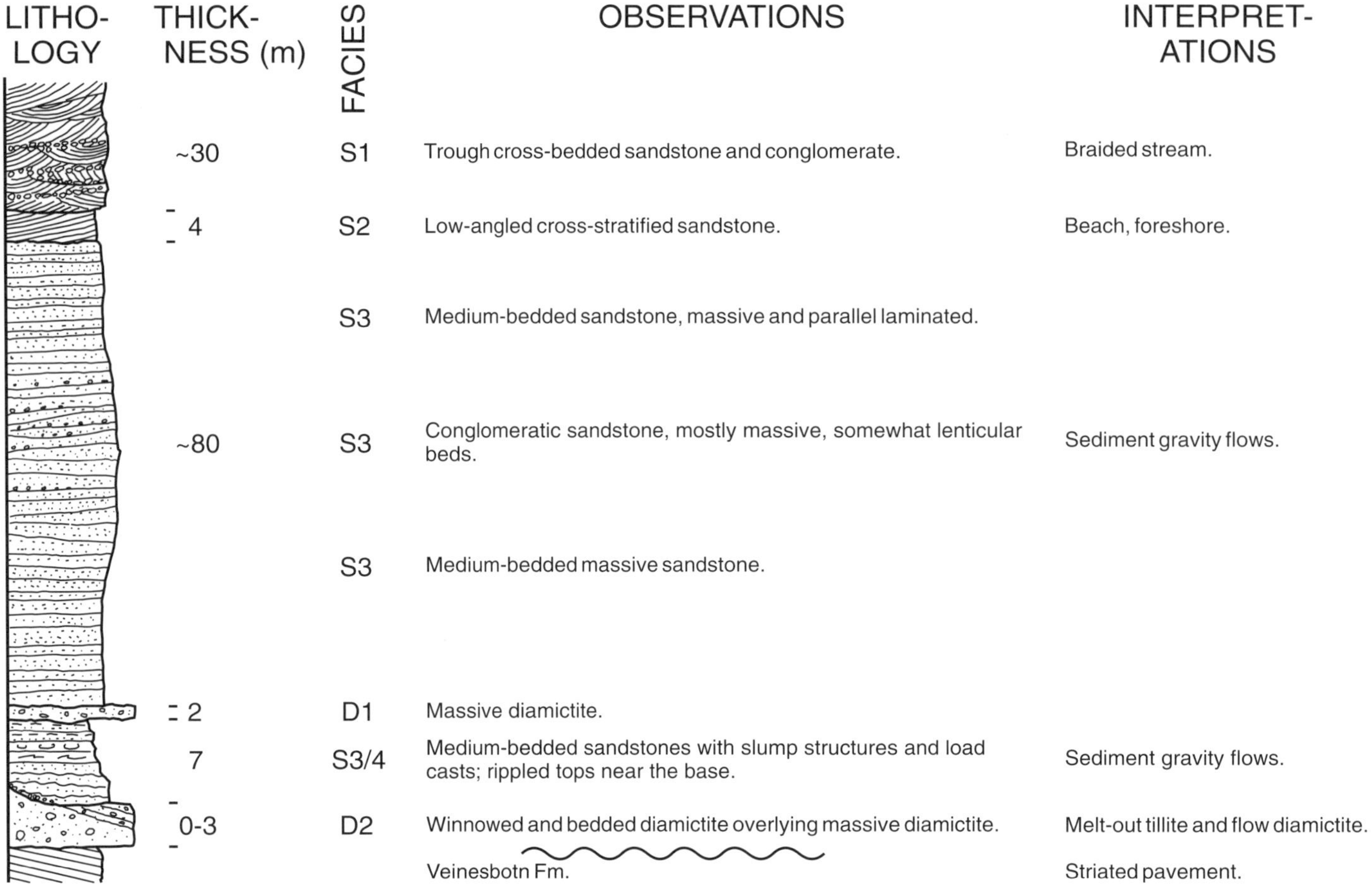

LITHO-LOGY	THICKNESS (m)	FACIES	OBSERVATIONS	INTERPRETATIONS
	~30	S1	Trough cross-bedded sandstone and conglomerate.	Braided stream.
	4	S2	Low-angled cross-stratified sandstone.	Beach, foreshore.
		S3	Medium-bedded sandstone, massive and parallel laminated.	
	~80	S3	Conglomeratic sandstone, mostly massive, somewhat lenticular beds.	Sediment gravity flows.
		S3	Medium-bedded massive sandstone.	
	2	D1	Massive diamictite.	
	7	S3/4	Medium-bedded sandstones with slump structures and load casts; rippled tops near the base.	Sediment gravity flows.
	0–3	D2	Winnowed and bedded diamictite overlying massive diamictite.	Melt-out tillite and flow diamictite.
			Veinesbotn Fm.	Striated pavement.

Figure 18. Stop 2.2. Facies within the Smalfjord Formation at Oaibaččannjar'ga (after Edwards, 1984).

correlated with the preglacial sandstones by Rosendahl (1931), but Føyn (1960), Crowell (1964), and Harland (1964) all assigned them to the Smalfjord Formation. Bjørlykke (1967) placed them back within the preglacial sequence (Veinesbotn Formation; Fig. 4), where they have remained, except in the recent work of Jensen and Wulff-Pedersen (1996), who again proposed placing them with the Smalfjord Formation.

Some beds of sandstone in the Veinesbotn Formation at Vieranjar'ga, on the south coast of Karlebotn and also some of the sandstones below the striated platform at Oaibaččannjar'ga contain abundant irregular, angular, and bedding-parallel oblate impressions (Fig. 19A); investigations at Vieranjar'ga revealed that these were weathered-out mud flakes. Similar, but partially worn-down, voids occur on the striated surface (Fig. 19B). Rice and Hofmann (2000) ascribed these also to weathered-out mud flakes; Laajoki (2002) disagreed. However, just northeast of the point where the main exposed platform comes to an end, extant mud flakes can be seen a few centimeters under the striated surface (Fig. 19C). In many cases, small feldspar grains or, less often, granitoid clasts can be seen in the mud-flake impressions, giving further evidence of a higher-energy regime.

Beneath the overhanging diamictite at the northeast end of the main striated platform, which lies by the coast at the southwest margin of the diamictite, there is a 2–3-cm-sized black clast lying within a depression in the pavement and also partially above the level of the pavement (Fig. 19D). This has caused enormous problems in interpreting the outcrop. Jensen and Wulff-Pedersen (1996) noted correctly that the clast fits perfectly into the depression; they interpreted this to indicate that the clast was pushed into the sediment and thus that the sediments were unconsolidated at the time of striation formation. This was a fundamental reason leading Jensen and Wulff-Pedersen (1996) to re-propose that the sediments were of Smalfjord Formation age and that the striations formed during soft-sediment deformation below a debris flow, now the overlying diamictite. Rice and Hofmann (2000), following Bjørlykke (1967) and Edwards (1975), regarded the underlying sediments as being much older and suggested that the black pebble was a lonestone, transported by high-energy currents comparable to those that formed the mud-flake conglomerates. In support of this, a 1-cm-sized, basement-derived, granitoid lonestone can be seen fully embedded within the striated platform near the black pebble (Fig. 19E). Another

Figure 19. Oaibaččannjar'ga. All at Stop 2.2. Smalfjord Formation overlying Veinesbotn Formation. (A) Imprints of mud-flake clasts on a bedding plane in the Veinesbotn Formation below the striated platform. (B) Striations at the NE corner of the main striated platform, near the sea, some with polished material (arrowed). Note depressions on the surface, interpreted as molds of mud flakes or other lonestones. (C) Greenish, distorted mud flake (arrowed) preserved within Veinesbotn Formation. (D) Black clast perfectly molded within the striated platform, under the overhang at the NE end of the striated platform. (E) Granitoid and other small lonestones (arrowed) within the striated pavement, under the same overhang as D. (F) White sandstone lonestone (or concretion?) within the Veinesbotn Formation.

possible lonestone, 4.5 by 3.5 cm in size, of sandstone, has been found at the NE end of the main polished platform, on a bedding surface a few centimeters below the striated surface (Fig. 19F), although this might be just a sandstone concretion. Similar-sized lonestones have been found elsewhere in the Veinesbotn Formation (Stop 3.4, Location 1).

The main platform carries numerous striations, with a number of dominant orientations (Fig. 20). Several of these have a smooth, semitranslucent polished texture (Fig. 19B). In thin section, this constitutes a very thin cataclasite with a cataclastic foliation parallel to the pavement surface, underlain by a thin layer of protocataclastic sandstone (Rice and Hofmann, 2000; Bestmann et al., 2006).

Under the scanning electron microscope with cathodoluminescence and electron backscattered diffraction (EBSD), the breakdown of detrital clasts to smaller fragments by fracturing and rotation can be seen in the protocataclasite. In the transmission electron microscope (TEM), abundant unstrained 120° triple grain boundary junctions have been observed in the foliated cataclasite. Bestmann et al. (2006) proposed that these could have formed either by solidification from a melt or by instantaneous recrystallization of highly strained comminuted material. Although both mechanisms require very high temperatures (with a 1-km-thick glacier and a slip rate of 0.2 m s^{-1}, typical of glacial earthquakes, flash temperatures of >1700 °C could have been reached), comparisons with fault rocks elsewhere indicates that the latter mechanism is more likely (Bestmann, 2008, personal commun.).

In the outcrops by the sea (where less recent erosion has occurred), just northeast of the main platform, it is evident that much of the striated surface is covered with a similar polish and that many striations have marginal ridges of breccia (arrowed, Fig. 21A). This breccia is paleorock flour, the source of loess. Note that within the polished striation of Figure 21A, there are three lateral ridges normal to the striation direction (arrowed); for the middle and right-side normal ridges, the polished surface seems to rise up gradually to the top and then rapidly fall away. The left-side normal ridge has a steeper profile and forms a shadow. The origin of these normal ridges is uncertain, but it is tempting to ascribe them to the end stage of different phases of stick-slip glacier movement.

Preservation of the polish, and more especially of the lateral ridges, implies that the ice did not move far after the striations formed. This is consistent with the model of dead ice for the deposition of the overlying diamictite proposed by Edwards (1975). The diamictite body is mound shaped, 3 m high and 70 m long (NE-SW), medium to darkish gray, unsorted, with 10%–20% clasts, consisting in part of rounded basement and sandstone lithologies, some from the underlying Veinesbotn Formation (Edwards, 1975). The clasts show several preferred orientations (Fig. 20), few of which are comparable with the underlying striation directions; from this, von Gaertner (1943) suggested that the clast and striation orientations were unrelated. Some clasts are faceted and striated (Bjørlykke, 1967).

On the main exposure of the diamictite seen on arrival at the outcrop, an obvious paler-gray vertical sedimentary sheet cuts the diamictite with remarkably sharp contacts (Fig. 21B). The clast composition and size range differ within the sheet, being finer grained and having fewer dark sandstone clasts. Similar vertical and horizontal sheets can be seen elsewhere, including on the upper surface of the diamictite. These are here presumed to be dewatering structures. This may have been a downward flow, due to subglacial fluid overpressures resulting from an overlying impermeable layer, possibly a sheet of ice. This forced the fluid to move downward, and then laterally, before escaping upward further away (see also Stop 6.2, Location 2, and Stop 7.1, Location 15). The example in Figure 21B is geometrically comparable to a mode 1 fracture with the initial stages of a lateral bridge structure forming at the base. Although this could be interpreted as an ice wedge, the similar rocks that formed horizontal layers/intrusions cannot be so interpreted.

The top of the central part of the diamictite is overlain by small pockets of lag conglomerate. Similarly, the margins of the diamictite, especially on the NE side, are overlain by pebbly, mostly massive, medium-bedded pinkish sandstones with prominent internal load structures and straight crested ripples. Intercalated thin, dark siltstones within these sandstones rapidly increase in thickness and frequency upward. Beds both onlap and are draped over the diamictite; the former is more common lower down, while the latter occurs higher up. Beds thicken rapidly away from the diamictite (Fig. 21C). These sediments were reworked from the diamictite after it had melted out from the ice and slumped outward (flow diamictite; Fig. 18).

To the northeast of the diamictite, at beach level, a number of other features can be seen in the Smalfjord Formation. Several

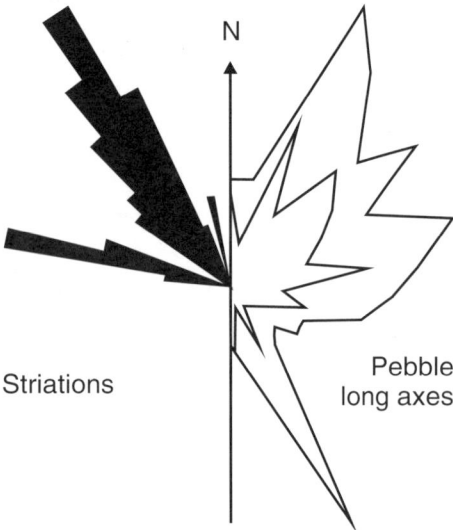

Figure 20. Stop 2.2. Rose diagram of orientations of striations (after Rice and Hofmann, 2000) and of pebble long-axes in the diamictite at Oaibaččannjar'ga (after von Gærtner, 1943).

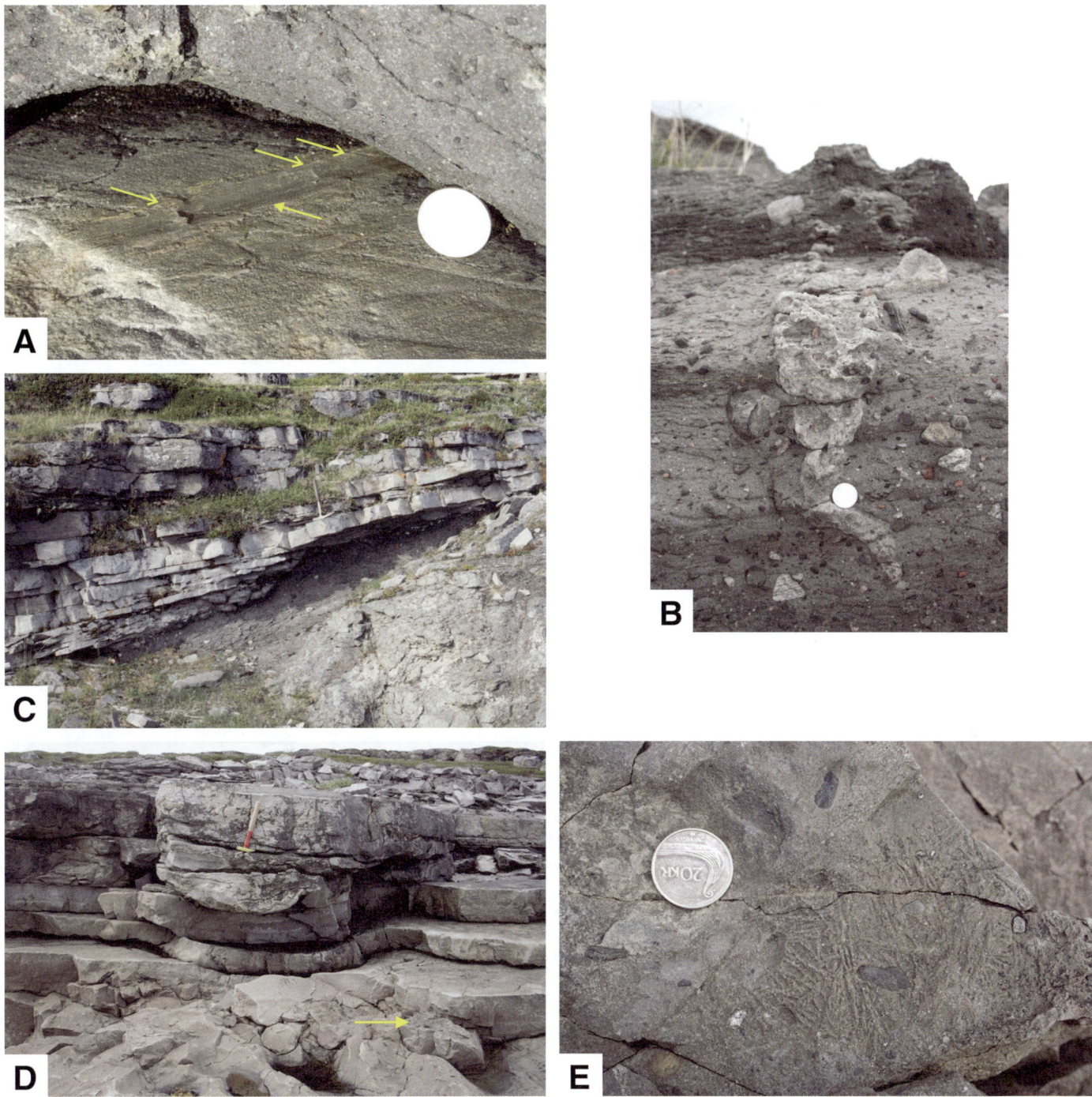

Figure 21. Oaibaččannjar'ga. All at Stop 2.2. Smalfjord Formation overlying Veinesbotn Formation. (A) Ridges composed of brecciated sandstone formed during polishing, parallel to and at the margin of the striation (solid arrow) and three ridges perpendicular to the striation direction (normal ridges; open arrows) in the Veinesbotn Formation; exposure just NE of the corner of the main platform. (B) Presumed fluid-escape structure in typical Oaibaččannjar'ga diamictite, caused by downward flow of overpressured water. (C) Top of diamictite, on the southwest side, showing complex onlap and draping of sediment of facies S3 onto the irregular upper surface of the diamictite. Onlap is more common in the lower part of the succession and draping is more common in the upper part, but onlap also occurs within the latter. (D) Bowl structure, interpreted as a small kettle hole. Note small patch of diamictite (arrowed; there are others present). (E) Possible molds of ice crystal structures. D and E are from the coastal outcrops NE of the main diamictite body.

small (centimeter-sized) pockets of diamictite occur (Edwards, 1975); the largest lies ~7 m northeast of the northeast tip of the main diamictite outcrop. These pockets all have the same appearance as the main diamictite. A further ~7 m northeast, a bowl-shaped structure, ~1.5 m diameter, is present in the Smalfjord Formation. The beds that bend down into the bowl retain a constant thickness, while the overlying beds thicken into the bowl, filling it (Fig. 21D). This might be a small ice-collapse structure (kettle-hole), filled by sand after ice melting. A further 2 m northeast, and lying approximately three bedding-planes above the present level of the beach cobbles, there are ridges and marks in the sandstones that might be molds of ice crystals (Fig. 21E) and, a further 11 m northeast, overturned blocks showing the 3-D structure of load casts lie on and near bedding planes in the Smalfjord Formation, with slightly asymmetric linguoid ripples indicating flow to the east. Occasional lonestones are present in the Smalfjord Formation sandstones here.

The overall evolution of the outcrop was documented in detail by Edwards (1975) and is shown schematically in Figure 22. Sediment-laden ice moved rapidly over the Veinesbotn Formation, forming the striations and associated high-temperature cataclasite. The last time this occurred, a part of the glacier remained as a block of subaerial dead ice after the rest of the ice melted and the associated sediments had been transported away (Fig. 22A). As the dead ice melted, sediment accumulated and some parts either became unstable, slumping down the sides of the till, burying the striated platform (Fig. 21E) and forming the draped layering of the flow diamictite, or were winnowed to form the lag conglomerate above the diamictite. Part of the melt-out till was also eroded, as a consequence of which the polish and ridges around underlying striations were eroded. Where not eroded, possibly because it was partly frozen, the mound of melt-out and flow tillite was buried by transgressive draping marginal marine sediments of the Smalfjord Formation, preserving the glacial polish.

In the Smalfjord Formation sediments above and slightly to the northeast of the main diamictite, well-developed ball-and-pillow structures and recumbent soft-sediment folds can be seen.

Stop 2.3: Larajœg'gi and Unnamed Hill

Location. Drive back to Mar'kan (Karlebotn village; Stop 1.3; Fig. 12) and, after asking for permission, walk north, climbing up the steep scarp to the crest of the old sea cliffs. From there, walk across relatively easy ground, locally boggy, to the obvious ridge (Larajœg'gi) ~750 m to the north. After inspecting the outcrops here, walk ~400 m N to NNW to the obvious, but unnamed, conical hill.

Grid ref. NT 5940 8175 and then 5935 8215, 1:50,000 map sheet Varangerbotn 2335 III, Edition 3-NOR. GEC 70°08′01″N, 28°33′57″E and then 70°08′14″N, 28°33′48″E.

Introduction. Four inliers of basement rocks occur within the Smalfjord Formation, in a zone trending NNW from Stop 1.4 and narrowing in that direction, to close to the road near Aune (Fig. 3). The most SSE of these has been studied in detail by Laajoki (2003), and this and all the other outcrops have been visited in detail by the authors. At all of these outcrops, sandstone beds of the Smalfjord Formation onlap the basement and, where the outcrop is good enough, can be seen to thicken slightly and dip gently away from the basement-cover contact. Outcrops of diamictite, with basement-derived clasts, are less common.

Fold axes in the area around Karlebotn trend approximately W-E to WNW-ESE (seen at Stop 2.1); such deformation is pervasive in the Nyborg Formation in the boggy area west of the road from Varangerbotn to Karlebotn. This trend is parallel to the large-scale, very open, essentially upright folds in the Smalfjord Formation in this area and to the long axis of the elongate basement inliers within the Smalfjord Formation at this and Stop 1.3. This allows two possibilities; most obviously, that the sediments in the Smalfjord Formation are draped over a shallowly buried, glacially formed undulating basement surface, passively forming E-W–trending folds; less obviously, that the folds and the basement inliers are *in part* hanging-wall anticlines related to minor displacement on a blind Caledonian thrust system developed in the basement, possibly as a passive-roof duplex with a relative back-thrust (top-to-N) thrust sense. Although the onlapping nature of the surrounding sediments makes the latter option seem unlikely, two points need to be remembered. First, the base of the strongly folded (shortened) postglacial Nyborg Formation to the west lies well below the top of the valley-infilling Smalfjord Formation, opening up the possibility of Caledonian tectonic uplift of the Smalfjord Formation in this area. Second, the hill Ruos'soai'vi, which lies to the west of the extensive boggy Nyborg Formation–filled area (Fig. 3; GEC 70°08′31″N, 28°16′13″E), is composed of allochthonous basement (cap dolostones dip 30°S on the south side of the hill, while bedding has been rotated to the vertical on

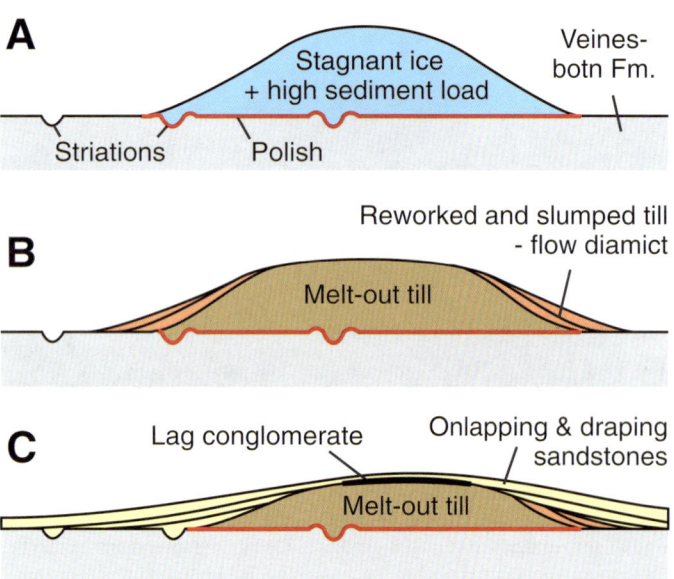

Figure 22. Stop 2.2. Schematic model for the evolution of the Oaibaččannjar'ga outcrop. See text for details. Note how the area in which polished material is preserved gradually decreases with time.

the east side and overall dips northward on the north side). If the folds on Selešnjar'ga are *not* tectonic in origin and there has been no thrust-related uplift, then there must be a significant approximately N-S–trending lateral ramp accommodating several kilometers of movement, essentially under the road south of Varangerbotn village, between the Smalfjord Formation on Selešnjar'ga and the Nyborg Formation to the west (see Introduction to Stop 2.1). No structural evidence for such a ramp has been recorded to date.

Description. The Larajœg'gi outcrop consists of a locally steep basement surface at the west end with a gentle slope toward the east, suggesting a roche moutonnée structure (cf. Laajoki, 2003) formed during westward Neoproterozoic glacial flow. On the slope at the west end of the outcrop, red, basement-derived diamictite can be found within hollows and fractures in the basement and as patches of sediment up to 20 cm thick. The sharp basement-diamictite contacts suggest ice plucking of the basement by the west-moving Neoproterozoic ice. Elsewhere, the smooth and fresh basement surface under the Smalfjord Formation supports the idea of erosion by ice, with rapid deposition of the sandstone afterward. Small outcrops of similar diamictite can be found on the north side of the basement outcrop, one in an underhang within the basement. Sandstones and quartzites of the Smalfjord Formation (likely facies S3; Fig. 18) onlap and thin toward the exposed basement, indicating it was a positive topographic feature at the time of deposition. Laajoki (2003) reported Neoproterozoic striations indicating westward-directed ice flow here, but these are more likely Pleistocene structures.

At the conical hill to the N/NNW, coarse sandstones with larger clasts (up to 15 cm size) and with some greenish-gray silts crop out in the southeast. About 70 m to the northwest, a sharp, 25-cm-deep notch in the basement is infilled with Smalfjord Formation sandstones, folded into a gentle syncline (Fig. 16C). Although this may be partially due to compaction of the sediments filling the depression, it dies away so rapidly upward that a primary, syndepositional component to the "folding" must also have been present. Such structures are typical for the Smalfjord Formation sediments infilling minor basement irregularities in the Karlebotn area. At the northern margin of the basement outcrop, there is a large exposure of coarse-grained sandstone with basement clasts up to 10 cm in size.

Day 3. Lower Smalfjord Formation on Skjåholmen and Vieranjar'ga, Varanger Paleovalley (III)

Introduction

The outcrops examined during this day lie at the base of the Smalfjord Formation (base Varanger paleovalley) on Skjåholmen, a small island in Varangerfjord, and at the western end of Vieranjar'ga (Veines), a peninsula jutting out from the south side of Varangerfjord (Figs. 3 and 23); in both cases, the unconformity lies within the Veinesbotn Formation. The outcrops are reached by boats taken from the Grasbakken area (Fig. 23).

The day starts with a cruise along the southern coast of Skjåholmen, pausing to take photographs and discuss the geology from the boats. A short landing is then made at the eastern end of the island, to examine the lithologies in detail. From there, the boats are taken to the northeast corner of Vieranjar'ga. After viewing the geology from the boats, the excursion lands and walks around to the southeast corner of the peninsula, examining the varying sedimentary facies on the way. The boats should be met at a pre-arranged point for the return trip to Grasbakken. Arrangements must also be made for the boats to collect the participants immediately if the sea conditions start to turn bad.

Remember to wear life jackets, properly fastened, at all times when in the boats.

Stop 3.1: Southeast Coast of Skjåholmen Viewed from the Boat

Location. Profile of the southeastern 700 m of the coast of Skjåholmen, examined by boat.

Grid ref. NT 6750 8175, 1:50,000 map sheet Varangerbotn 2335 III, Edition 3-NOR. GEC 70°08′86″N, 28°48′34″E and to the east.

Introduction. This profile exposes the contact between the shallow-marine Veinesbotn Formation and various facies of the Smalfjord Formation (Figs. 24 and 25). Along the profile, note the irregular contact geometry and the facies changes that can be readily observed in the Smalfjord Formation; it is important to remember that this contact is a part of the regional angular unconformity at the base of the Varanger paleovalley.

Description. The traverse starts at the most westerly outcrop of the Veinesbotn Formation, exposed at a vertical paleoscarp

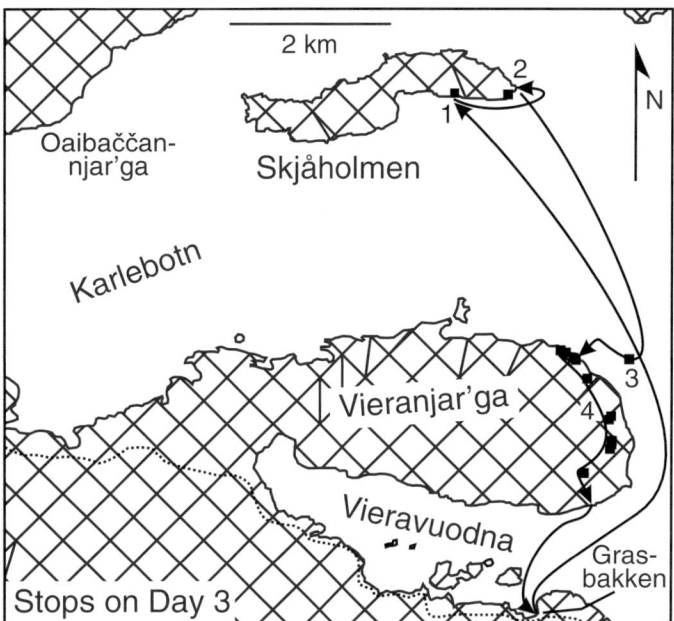

Figure 23. Map showing the location of outcrops visited during Day 3. Dashed line indicates roads.

LITHO-LOGY	THICKNESS (m)	FACIES	OBSERVATIONS	INTERPRETATIONS
	30	S1	Intercalated sandstone and conglomerate. Poorly sorted conglomerate in channels up to 3 m high. Trough cross-bedded sandstones up to 1 m high and parallel-laminated sandstones in gently inclined beds forming units up to 4 m high. Many sets have basal conglomerates and occasional soft-sediment folds.	Braided stream deposits formed by channel migration and filling and by bar migration. Flow was to the west and northwest.
	10	S2	Large-scale foresets, dipping at 5-10° to the west. Consist of parallel-laminated brown sandstone with scattered pebbles and conglomerate lenses. Pebbles of dolomite, crystallines and sandstone.	Delta foreset that prograded toward the west.
	0-4	D1	Mostly massive tillite; to the west, it is sandy with dolomite clasts, and to the east, it is crystalline and much sediment is derived from underlying unit. Both tillites have local clast-rich zone at top, up to 1 m thick.	Probably lodgment tillite, suggested by erosion and incorporation of underlying sediment. Possible subaerial weathering and winnowing at the top.
	0-15	S4	Thin- to medium-bedded, parallel-laminated sandstone with intercalated mudstone. Sharp contacts, occasional grading, and erosive, pebbly bases. Isolated gravel lenses, and pebbles scattered in sandstone and mudstone. Abundant soft-sediment folding with axial planes tipped to the west. Clasts of dolomite, sandstone and crystalline. Local basal breccia of large angular blocks of Veinesbotn Fm.	Sediment deposited rapidly on a slope by turbidity currents or sediment gravity flows. Basal breccia may be reworked glacial drift.
			Unconformity: local strong relief as near-vertical walls up to 4 m high, and striking N-S and NE-SW. Sharp breaks in slope.	Glacially scoured surface may have been locally modified by wave and current erosion. Steep slopes may be paleo-wave-cut cliffs.
			Veinesbotn Fm.; Vadsø Group.	

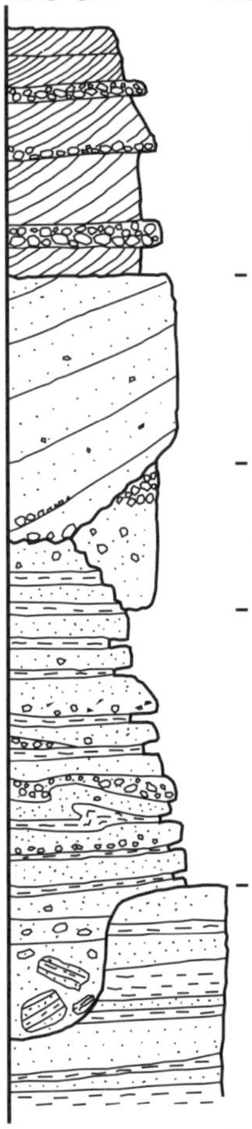

Figure 24. Stops 3.1 and 3.2. Facies within the Smalfjord Formation on Skjåholmen (after Edwards, 1984).

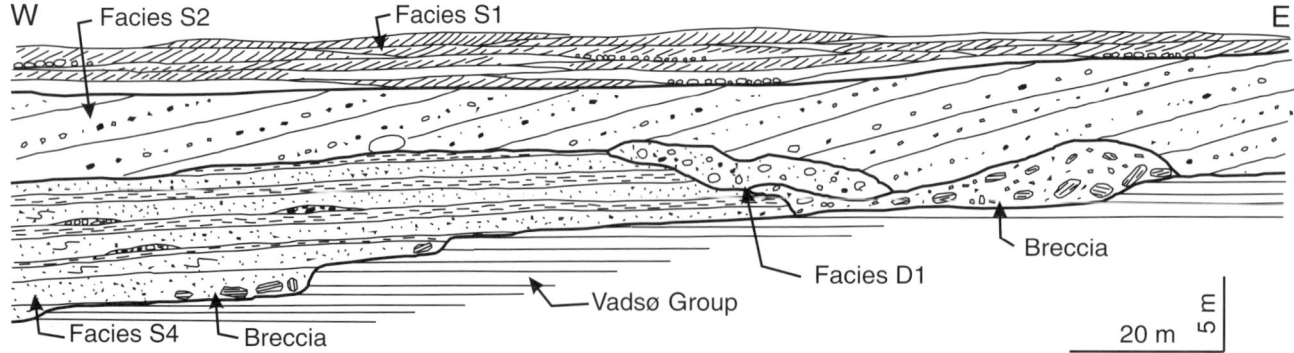

Figure 25. Stops 3.1 and 3.2. Semi-schematic profile along the south coast of Skjåholmen, showing the distribution of facies outlined in Figure 24 (after Edwards, 1984).

(Fig. 26A), ~700 m from the easternmost part of Skåjholmen; facies remain constant within this coastal section. Large jumbled blocks of the Veinesbotn Formation, fallen from the scarp, lie at the bottom of the Smalfjord Formation.

The heterolithic, well-bedded character of the Veinesbotn Formation here may partly explain why striated pavements have not been observed from this area; beds of material would have been more easily eroded than striated. Moving to the east, parallel to the shoreline, three contrasting facies (facies S4, D1, and S2; Figs. 24 and 25) can be observed in the lower part of the Smalfjord Formation. The lowermost facies consists of bedded sandstones and occasional conglomerates (Figs. 26A and 26B) that have been interpreted as subaqueous mass-flow deposits (facies S4; Edwards, 1984; Arnaud, 2008). These are overlain by sporadic diamictites (D1) that are apparently connected by an extensive erosion surface (cf. Figs. 24 and 25). Locally, the diamictites are associated with shear and deformation of the underlying sandstones. The diamictites cut through both the undeformed and the sheared sandstones (Fig. 26C). These diamictites were interpreted as subglacial till by Edwards (1984) and as sediment gravity-flows by Arnaud (2008). They are erosively overlain by a thick (~10 m), sandy clinoform set that dips toward the west. This has been interpreted as a prograding delta-front slope (facies S2; Edwards, 1984). The toe of these foresets, along with a large granite clast, probably an erosional remnant of a diamictite layer, is shown in Figure 26D, resting on a erosional contact on facies S4. Further up the slope, beyond close inspection, there are trough cross-bedded sandstones, interpreted as braided-stream deposits, indicating flow to the west and northwest (facies S1; Edwards, 1984).

Stop 3.2: Skjåholmen Eastern End

Location. Land at the east end of the island and walk due west above the southern coastal rocks for ~70 m and then go

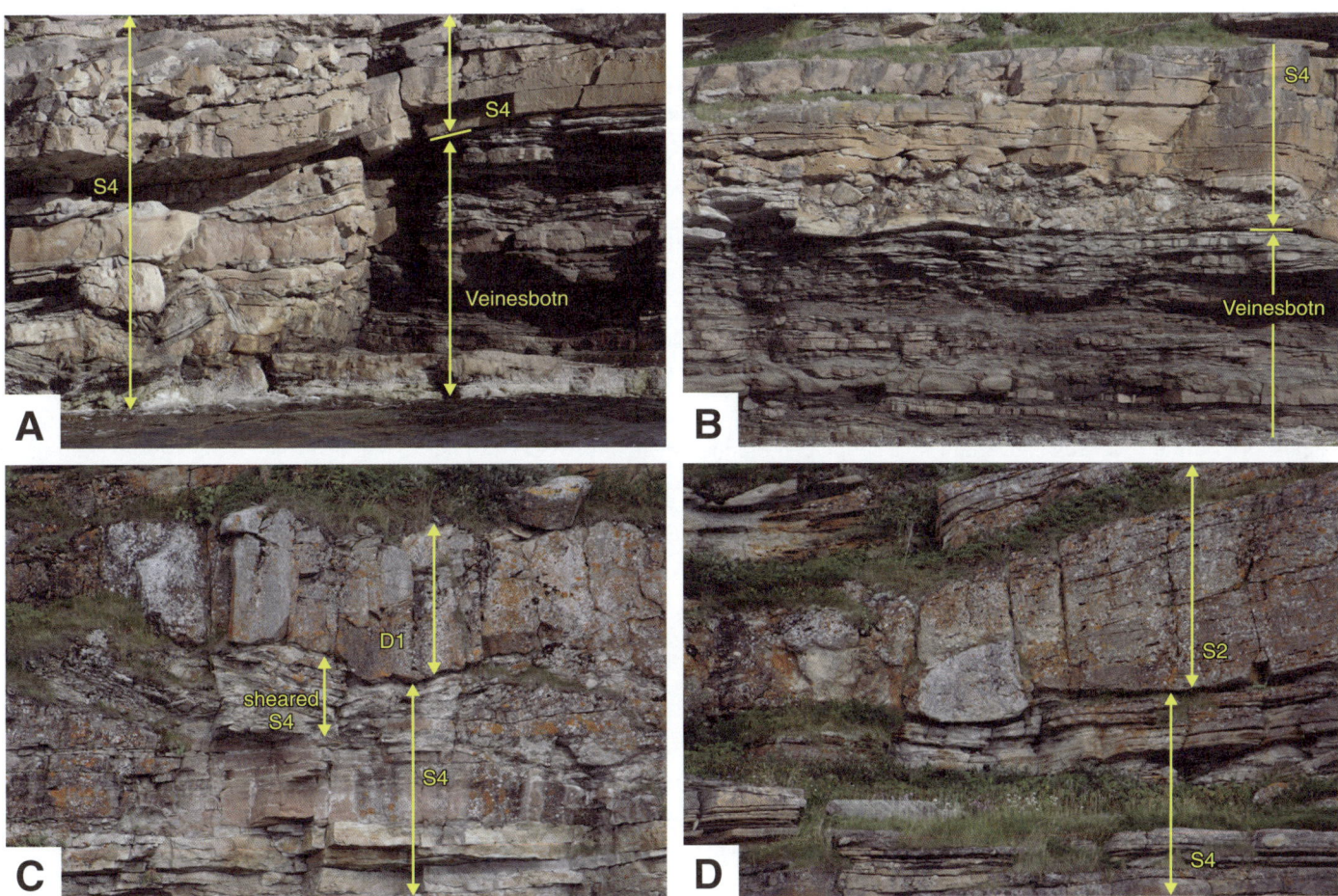

Figure 26. Coastal profile, Skjåholmen. Facies refer to Figures 24 and 25. All at Stop 3.1. (A) Vertical paleoscarp juxtaposing Smalfjord Formation facies S4 (left) against the Veinesbotn Formation (right). Note massive scarp debris on far left. (B) Subaqueous mass-flow deposits at the base of the Smalfjord Formation facies S4, overlying darker rocks of Veinesbotn Formation. (C) Horizontally bedded facies S4 at the base of the picture are truncated by glacially sheared S4 deposits formed during the deposition of the overlying massive diamictite of facies D1, which also cuts the sheared rocks. (D) Toeset of large west-dipping clinoforms, facies S2, erosively overlying bedded sandstone of facies S4. Along the erosion surface, there is a large granitoid clast, likely a remnant of glacially deposited diamictite of facies D1.

down to a small bay; its position can be noted from the boats shortly before landing. The boats should wait at the landing point; examining the outcrop should not take more than 1 hour.

Grid ref. NT 6875 8176, 1:50,000 map sheet Varangerbotn 2335 III, Edition 3-NOR. GPS 70°08′03.9″N, 28°48′53.3″E.

Introduction. The Smalfjord Formation infilling the Varanger paleovalley exhibits a wide range of facies, varying rapidly both laterally and vertically (Figs. 24 and 25), reflecting a progression of contrasting depositional environments interacting with the basal unconformity surface. Here, the units observed on the boat ride can be studied in more detail, as they are developed in an easily accessible, well-exposed outcrop.

Suggested key questions are: Is there any direct evidence for glacial deposition? Are the diamictites subglacial tillites or mass-flow deposits? And, are there any robust environmental indicators that everyone can agree about?

Description. The unconformity here between cross-bedded white to pale-gray sandstones of the Veinesbotn Formation and onlapping buff facies S4 sandstones of the Smalfjord Formation is a smooth curviplanar surface (Fig. 27A). Although there is no basal conglomerate, there are abundant clasts of the Veinesbotn Formation in the immediately overlying sandstone. The lower part of the overlying succession consists of up to several meters of sandy gravity-flow deposits (facies S4; Figs. 24 and 25) with ball-and-pillow and flaser bedding structures (Fig. 27B) and then by lenticular intraformational conglomerates (Fig. 27C), remnants of diamictites (facies D1), and clinoforms of facies S2.

Toward the eastern end of the bay, facies D1 is better developed (Fig. 27D). Here, facies S4, with ball-and-pillow structures, is erosively overlain by up to several meters of diamictite interpreted as either mass-flow deposits (Arnaud, 2008) or a tillite (Edwards, 1984). The basal part of this is an erosional remnant of a deformation diamictite, with large clasts of the Veinesbotn Formation. This is overlain by a subglacial diamictite with subrounded basement clasts. Most of this has a pale matrix, except at the top, where the clay fraction has apparently been winnowed out due to either reworking or, possibly, subaerial exposure. The clasts at the base of the diamictite, derived entirely from the Vadsø Group, are angular to rounded; shale clasts are deformed. The diamictite above (described as a boulder conglomerate by Arnaud, 2008) has large, well-rounded boulders of white basement granite, especially near the top, clearly derived from a more distal source.

Both the turbidites and the diamictites are overlain by westward-dipping coarse-grained, sandy delta-front clinoforms (facies S2) composed of rounded dolostone clasts in a quartz-rich gritty matrix (Fig. 27E, taken near the boat landing point). The clinoforms and the overlying sandstones and conglomerates (facies S1) record high-energy deltaic and glaciofluvial conditions (Edwards, 1984; Arnaud, 2008).

Stop 3.3: View of NE Corner of Vieranjar'ga Peninsula

Location. Return southward to Vieranjar'ga and stop offshore to examine the profile from the boat (Figs. 28A and 29).

Grid ref. NT 7030 7855, 1:50,000 map sheet Nesseby 2335 II, Edition 2-NOR. GEC 70°06′08″N, 28°51′05″E.

Introduction. This profile exposes the unconformable contact between the shallow-marine Veinesbotn Formation and the Smalfjord Formation, as well as the contact between facies S1 and S3 within the latter (Fig. 30).

Description. The panorama (Fig. 29) shows that the pinkish colored sandstones and dark-gray shales of the Veinesbotn Formation dip northward relative to the angular unconformity with the overlying Smalfjord Formation. This unconformity is part of the regional unconformity, also seen on days 1 and 2, forming the base of the Varanger paleovalley (Bjørlykke, 1967). Since the Smalfjord Formation deposits are currently essentially horizontal, the regional tilting of the Veinesbotn Formation must have occurred earlier (perhaps as a flanking structure in response to the postulated fault along Varangerfjord proposed by Røe, 2003). Within the Smalfjord Formation, facies D1 and D2 are cut out to the south of here (Fig. 28), and hence the orange-buff–colored cliffs comprise only facies S1 (Fig. 30), a complex succession of interbedded conglomerates and sandstones from a braided-stream environment, with a wide range of clast types (Edwards, 1975). The contact to the overlying white sandstones of facies S3 (Edwards, 1984; Fig. 30) lies directly above the cliff.

Stop 3.4: Vieranjar'ga Peninsula

Location. From the last stop, go toward the northwest and land on the northeast side of Vieranjar'ga (Fig. 28), at the west side of a gravelly beach to the west of which are the obvious well-bedded sediments seen in Figure 31A. Be sure to land on the correct beach. The first part of this stop should be visited at relatively low tide. Arrange with the boat owners to meet at the end of the profile, some 5 hours later. The persons with the boats should be asked to return earlier if they think the sea conditions will turn bad.

Grid ref. NT 6965 7876, 1:50,000 map sheet Nesseby 2335 II, Edition 2-NOR. GPS 70°06′15.3″N, 28°50′00.6″E.

Introduction. The section around the western end of Vieranjar'ga shows the geometry of the contact between the Smalfjord and underlying Veinesbotn Formation in 3-D, as well as the distribution of a variety of facies in the Smalfjord Formation (Figs. 28B, 31; Edwards, 1975, 1984; Arnaud and Eyles, 2002). As the excursion works through the section, participants should ponder on similar issues to those at Stop 3.2: Are there any deposits that offer a robust reference point in terms of depositional environment? Are the diamictites subglacial, proglacial, or glacial debris retransported in submarine mass flows? The first few locations are based in Unit X (facies S4, but deposited below facies S1; Edwards, 1975), while the later ones are in facies D1, D2, S1, and C1 (Edwards, 1975).

Description. *Location 1.* GPS 70°06′15.3″N, 28°50′00.6″E. At the landing site, alternating rust-spotted sandstones in beds up to 0.5 m thick, sometimes lensoid, are interlayered with dark silty rocks with thin-bedded sandstones (Fig. 31A). Some sandstone

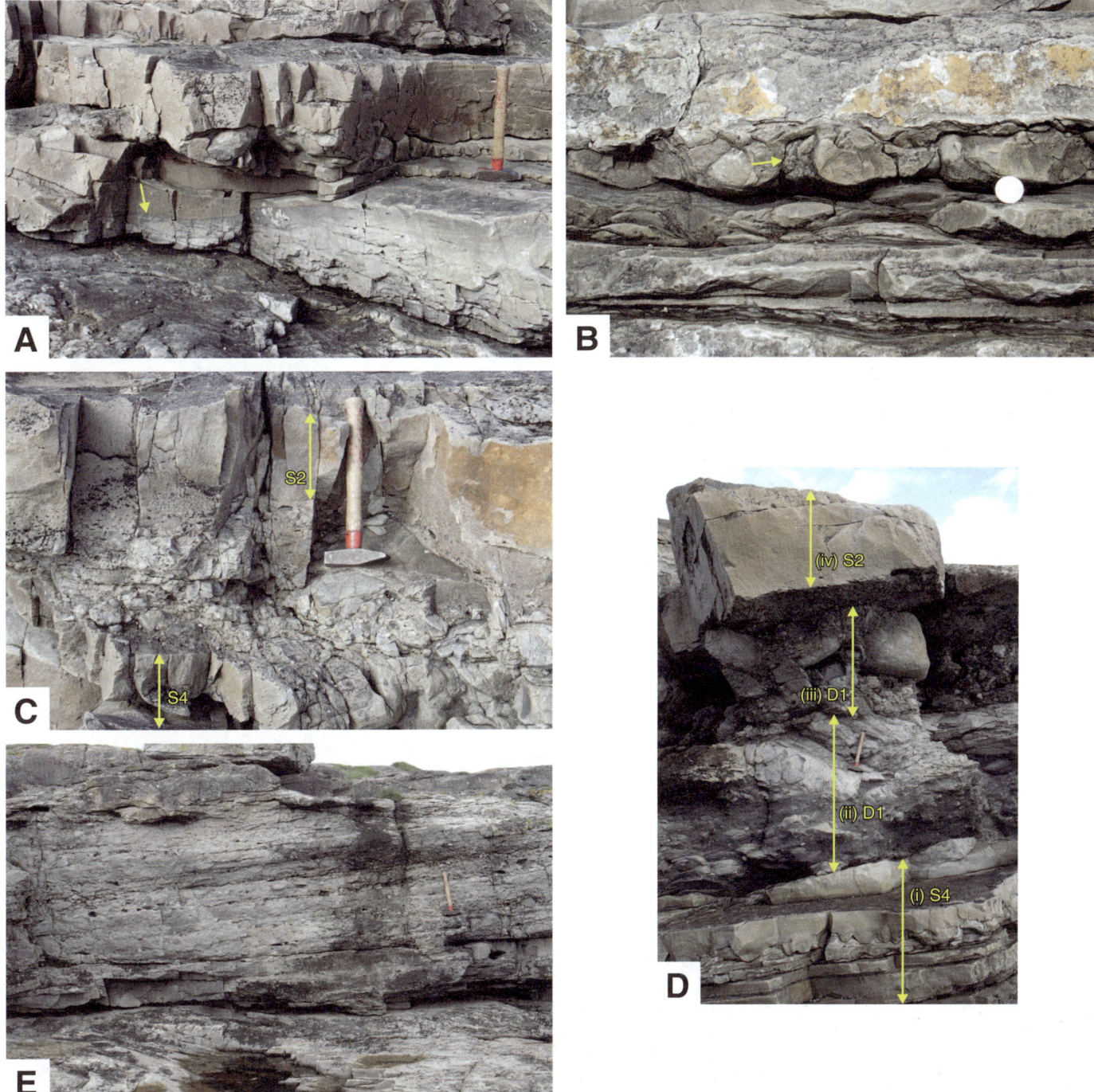

Figure 27. East Skjåholmen. Facies refer to Figures 24 and 25. All at Stop 3.2. (A) Regional angular unconformity (arrowed) between the underlying cross-bedded gray to white sandstones of the Veinesbotn Formation and the overlying onlapping buff-weathering sandstones of facies S4 in the Smalfjord Formation. Large randomly oriented clasts of Veinesbotn Formation lie toward the base of the Smalfjord Formation. (B) Ball-and-pillow structure in facies S4 sandstones, Smalfjord Formation. Note the breccia-like appearance of the rock between the balls (arrowed) and the lenticular bedding below the coin. (C) An erosional remnant (at the level of the hammer head) of diamictite facies D1, represented by an intraformational conglomerate. This is underlain by sandstones beds of facies S4 and overlain by sandstones of facies S2. (D) A vertical section made up of (from base to top): (i) ball-and-pillow sandstones of facies S4; (ii) gray deformation tillite with large clasts of Veinesbotn Formation sediments of facies D1; (iii) winnowed diamictite of facies D1, comprising subglacial tillite with abundant rounded basement clasts; and (iv) faintly stratified sandstones of the facies S2 clinoforms. (E) W-dipping coarse-grained delta-front clinoforms (facies S2) with dolostone clasts.

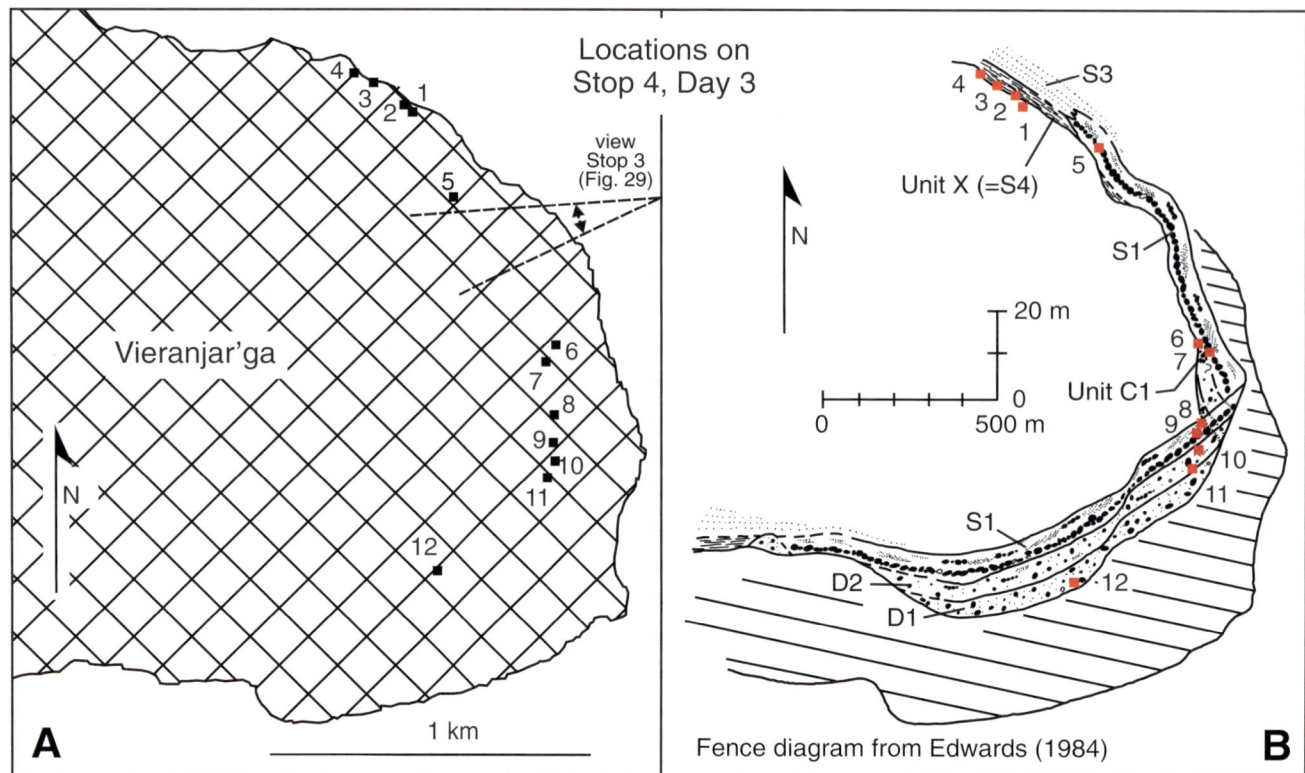

Figure 28. Map (A) and fence diagram (B) showing the location of outcrops visited during Stop 3.4 (after Edwards, 1984).

beds show bed forms and intraformational breccias or lonestones (Fig. 31B). Dark-purple iron-rich sandstone beds, up to 1 m thick, also occur. These sediments belong to the Veinesbotn Formation, at the base of the Vadsø Group. Although Røe (2003) suggested correlating the Veinesbotn Formation with the Gamasfjell Formation (Tanafjord Group; Fig. 4D), the presence of dark shales and ferruginous sandstones strongly suggests that a correlation with the slightly older Dakkovarre Formation is more likely, if any correlation with the Tanafjord Group is to be made. From here, walk westward along the coastal exposure.

Location 2. GPS 70°06′16.4″N, 28°49′58.9″E. The basal unconformity of the Smalfjord Formation is exposed here (Fig. 31C), overlain by graded conglomeratic sandstones deposited by sedimentary gravity-flows of Unit X.

Figure 29. East Vieranjar'ga panorama view. Stop 3.3. View of the unconformity between the underlying, north-dipping Veinesbotn Formation and the overlying Smalfjord Formation at the northeast end of Vieranjar'ga (see Fig. 28A for field of view). The cliff consists of buff-weathering facies S1 (Fig. 30). Facies S3 forms the white sandstones above the cliff. The small cairn mentioned in Stop 3.4, Location 5, is on the right side. Ladder ~2 m high.

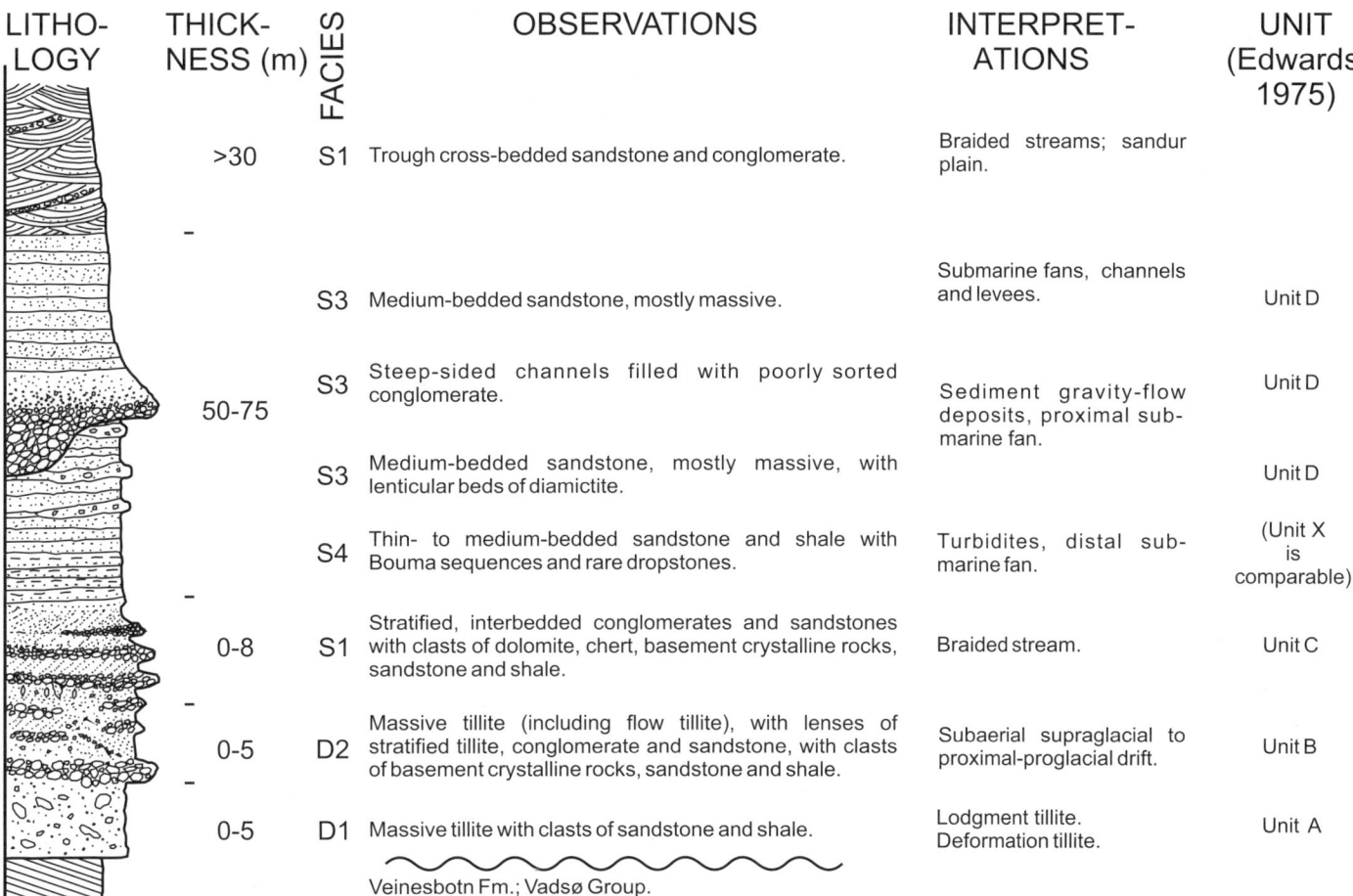

Figure 30. Stop 3.4. Facies within the Smalfjord Formation on Vieranjar'ga (from Edwards, 1984).

Location 3. GPS 70°06′18.1″N, 28°49′53.0″E. Slightly further west, down across a wave-cut platform with ripples and under an overhang (Fig. 32A), superb load casts (or pseudonodules, since they are detached from their original bed) can be seen near the base of the Smalfjord Formation (Fig. 31D); a polymict conglomerate forms the base of the Smalfjord Formation here.

Somewhat to the east of the overhang, an enigmatic white sandstone clast lies across the erosional contact between fine-grained green sandstones and coarser red sandstones, seemingly cutting the bedding in the former (Fig. 32B; location shown in Fig. 32A). This clast appears to have either been dropped in or rolled in with one of the gravity flows; the bedding below the clast is bent down and is folded adjacent to both sides of the clast (arrowed). On the east (left) side, both the green sandstones and an overlying package of thin-laminated siltstones and fine sandstones onlap onto the white clast. The overlying coarser-grained red sandstones onlap and drape the white nodule, although the top layer of the thin-laminated sequence seems to be contiguous with layering in the red sandstones. Curiously, the individual layers in this thin-laminated package thin laterally and downlap; this occurred during deposition and is not related to deposition of the red unit.

Note that the pavement in the Veinesbotn Formation to the north contains two sets of megaripples; one set, with wavelengths of ~2 m, is NNE-SSW oriented and gently asymmetric to the east, and the other, orthogonal set has wavelengths of ~0.75 m and is essentially symmetrical. Participants during the original field excursion noted that the ripple troughs in the former set are unconformably filled with glacial sediments, indicating they are neither tectonic nor soft-sediment folds.

The complex geology within Unit X in the cliff section behind the wave-cut platform (Fig. 32A) has not been studied in detail. This unit postdates deposition of facies D1 and D2 but may have been related to, but predates, deposition of facies S1 (Edwards, 1975). All beds seen in the cliff thin toward the east and are eventually cut out. Within this overall geometry, some beds are cut out at the base of the succession (the rippled, wave-cut platform of Veinesbotn Formation), while others are cut out higher up, within the cliff, along presumed erosion surfaces (arrowed). Although not properly analyzed, it appears to be a series of nested channels, successively eroding into older channel

Figure 31. East Vieranjar'ga. Facies refer to Figure 30. All at Stop 3.4. (A) Location 1. Dark-gray shales and pale-gray sandstone of the Veinesbotn Formation at the landing point. (B) Location 1. Intraformational breccia in Veinesbotn Formation. (C) Location 2. Graded, conglomeratic, buff-colored sandstones at the base of the Smalfjord Formation (Unit X), onlapping gray sandstones below the angular unconformity at the top of the Veinesbotn Formation. (D) Location 3. Flow-folds and load structures (pseudonodules) in sandstones of Unit X, close to the base of the Smalfjord Formation.

sediments, deposited as sedimentary gravity flows (turbidites; Edwards, 1975) in a very high-energy marine environment. The top of the cliff, only just visible in Figure 32A, consists of coarse-grained conglomerates within facies S3.

From here, walk around the headland, over the seaweed, to the next bay. On the way, climbing ripples can be seen in Unit X of the Smalfjord Formation (Fig. 32C).

Location 4. GPS 70°06′18.7″N, 28°49′49.6″E. Here, evidence of soft-sediment folding can be seen on both a large scale (facies S3) in the cliff (Fig. 32D) and also on a small scale, in the outcrop below (Unit X; Fig. 32E). The former indicates instability on a slope that was prograding toward the west, whereas the latter reflects shear from the overlying flow (Edwards, 1975, 1984).

Location 5. GPS 70°06′07.3″N, 28°50′13.3″E. From the last locality, climb the hill above the cliff and then walk toward the east-southeast. A large outcrop of orange-buff–weathered, poorly sorted conglomerates on the north side, 10 m or more high, contains sandstone lenses up to 50 cm thick and several meters long. This unit, which has a sharp erosive contact with the underlying sandstones of facies S3, was deposited in a large channel during sediment gravity flows (Fig. 30). From here, walk further southeast to the obvious small cairn and then either go directly downhill to the bottom of the Smalfjord Formation to the north of the view in Figure 29 and then walk along the base of the Smalfjord Formation to Location 6, or walk southward to the minor summit to the SSE and then go downhill to Location 7, through the lithostratigraphic profile documented at that locality.

Location 6. GPS 70°05′53.0″N, 28°50′44.0″E. This point marks the northern limit of Unit C1, the winnowed relict of facies D1 (Fig. 28). From here, walk southward along the base of the Smalfjord Formation outcrops to the next location.

Location 7. GPS 70°05′51.3″N, 28°50′40.8″E. This very significant outcrop shows well-developed cross-stratification that was interpreted by Edwards (1975, facies S1; Fig. 30) to indicate

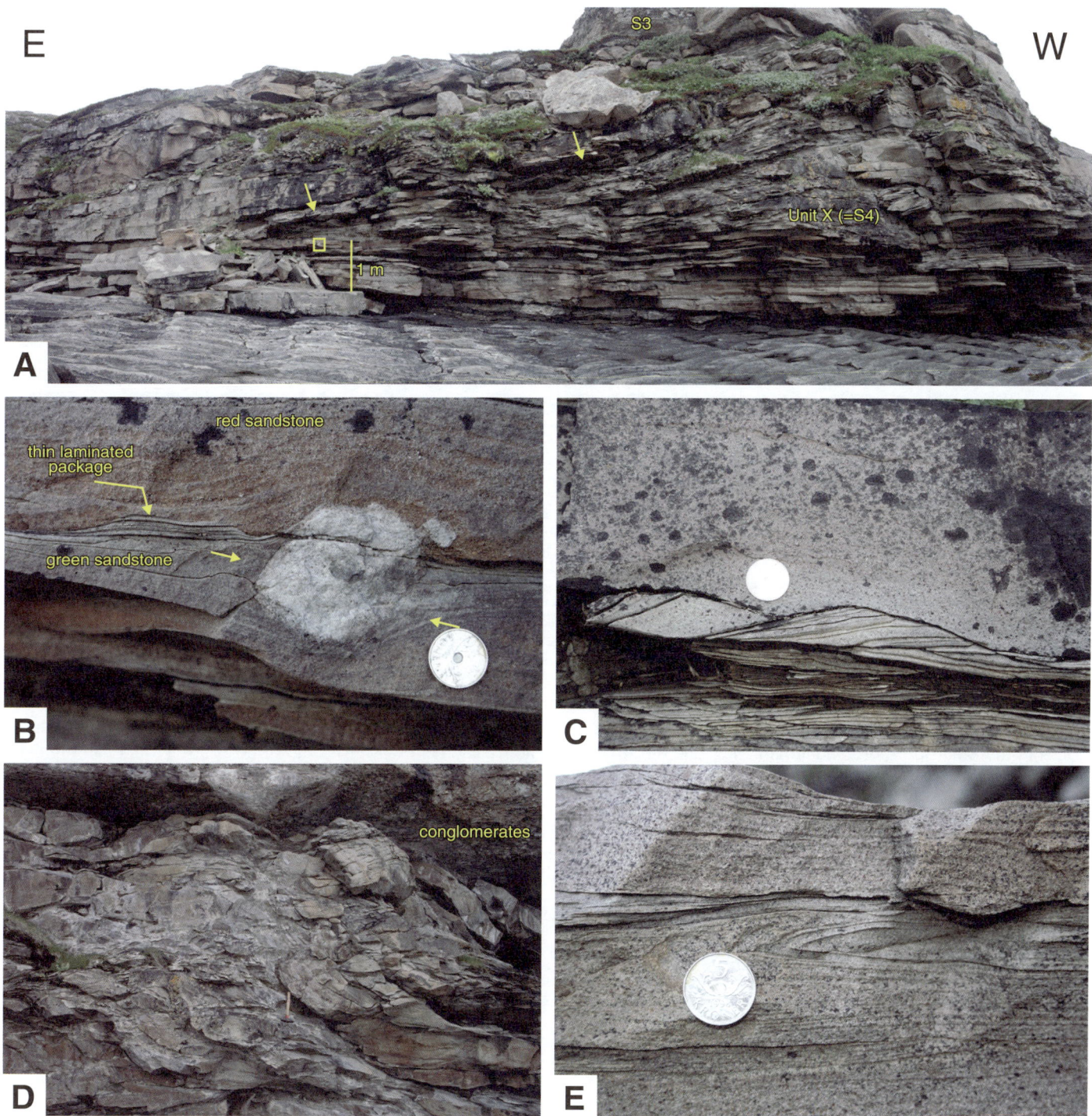

Figure 32. East Vieranjar'ga. All at Stop 3.4. All Smalfjord Formation except where stated otherwise. (A) Location 3. View of wave-cut platform in the Veinesbotn Formation and the cliff behind, showing a complex of nested channels in Unit X, in which beds thin to the east. See text for details. Location of Figure 31D is on the right, directly above the platform. Location of Figure 32B is indicated by yellow square. (B) Location 3. Enigmatic clast cutting and draped by sediments in facies Unit X; note soft-sediment folds (arrowed) adjacent to the clast. (C) Location 3. Climbing ripples in Unit X. (D) Location 4. Large-scale soft-sediment fold, overlain by massive conglomerates (cf. Locality 5); both within facies S3. (E) Location 4. Small-scale soft-sediment fold in Unit X.

fluvial deposition. (Note that the complete profile is shown from top to bottom in Figs. 33A, 33C, and 33E.)

The base of the succession consists of the remains of ~0.5 m of deformation tillite derived from the upper part of the Veinesbotn Formation. These are overlain by ~2.5 m of matrix-supported conglomerate (Unit C1; Fig. 33E), with clasts derived by winnowing of facies D1 prior to the deposition of the overlying facies S1, which includes both stratified conglomerates and sandstones filling low-angle channels. The conglomerates show sorting and parallel orientation of (?intraformational) platy clasts (Fig. 33C).

These are overlain by orange-colored conglomerates and sandstones (Fig. 33A), forming the top of the cliff section that delineates the base of the Smalfjord Formation (top of facies S1; Fig. 30). Above the cliff section, these are overlain by >20 m of pale-weathering, coarse- to medium-grained, parallel-laminated sandstone with few sedimentary structures (some load casts), forming the gentler hillslope above (S3; Figs. 29 and 30). The top of the hillside profile consists of 0.5 m of orange (dolomitic) sandstone with common rounded lonestones and then by 0.5 m of conglomerate with well-rounded clasts.

From the base of the profile, walk almost due south, across the steep valley.

Location 8. GPS 70°05′44.1″N, 28°50′44.5″E. This outcrop, found by Tony Spencer during the original excursion, shows several subvertical sandstone sheets up to 20 cm wide and trending approximately E-W, within the Veinesbotn Formation (Fig. 33B). One of the fillings broadens out at the base and may have joined with another sheet out of the plane of view.

These were initially interpreted as sandstone wedges. However, the coherent angular clasts of Veinesbotn Formation sandstones in the Smalfjord Formation diamictites and the striations at Oaibaččannjar'ga (Stop 2.2) both indicate that the Veinesbotn Formation was lithified prior to the onset of glaciation, and hence sandstone wedges could not have formed.

Two alternative models can be proposed. First, the sandstone sheets are fracture fillings between blocks of Veinesbotn Formation that had been slightly displaced, perhaps by a freeze-thaw mechanism or by glacier movements. This seems rather unlikely since diamictites (facies D1) form the basal unit of the Smalfjord Formation on Vieranjar'ga (Edwards, 1975, 1984). Second, they are sandstone dikes, derived from below.

Location 9. GPS 70°05′43.3″N, 28°50′43.6″E. The angular unconformity between the Veinesbotn Formation and the Smalfjord Formation is here marked by a thin sedimentary breccia.

Some 20 m further south, an ~4-m-thick diamictite lies at the base of the succession. The basal part is a deformation diamictite with clasts and matrix derived from the Veinesbotn Formation (facies D1). This is overlain by ~1.5 m of matrix-supported diamictite with exotic (basement-derived) clasts, similar to the diamictite at Oaibaččannjar'ga (D2). Within this upper diamictite, there are also purple sandstone clasts up to several meters long and generally <25 cm thick (Fig. 33D), although some clasts are larger and themselves contain lonestones. This diamictite is overlain by ~1 m of conglomerate, perhaps clast supported (Fig. 33F), and then more diamictite (still facies D2). Conglomerates, sometimes showing clast imbrication (Fig. 33G, arrowed), above this diamictite have sandstone beds and some sorting (S1); the sandstones are stratified in part. This succession was deposited during large fluctuations in current strength.

Differentiation of facies D2 and S1 here is subtle rather than obvious. Facies S1 has low-angle cross-bedding throughout, reflecting its high-energy depositional environment, and typically weathers to form continuous subvertical outcrop faces. In contrast, facies D2 consists of alternating massive diamictites and conglomerates/sandstones with cross-bedding. The former weather more easily and so are eroded out, leaving a series of ledges of conglomerates or sandstones.

Location 10. GPS 70°05′ 41.4″N, 28°50′44.3″E. This outcrop shows a profile through a complex channel in facies D2. The channel is eroded into matrix-supported diamictite (base of Figs. 34A and 34B). The initial channel fill has a preserved minimum relief of ~1 m and consists of white sandstones with some very thin conglomerates (Fig. 34B; below the hammer). A further phase of erosion was followed by cyclical deposition of conglomerates and cross-bedded sandstones; at least four channels can be identified in Figures 34A and 34B.

Location 11. GPS 70°05′39.8″N, 28°50′42.3″E. A similar channel within diamictites of facies D2 is seen in Figure 34C. The diamictite predominantly contains siliciclastic sedimentary clasts, although larger rounded basement-derived clasts are also present (Fig. 34D). The channel was filled with cross-bedded white sandstones (Fig. 34C) that are overlain by conglomerates on an erosional surface. In more detail, note the wedge of tillite (?flow tillite) at the bottom left of Figure 34C, which is bounded above and below by erosional surfaces also (1 and 2 in Fig. 34C, with 3 being the erosional surface at the base of the conglomerates).

Location 12. GPS 70°05′30.2″N, 28°50′10.3″E. A large megaclast of red Veinesbotn Formation sandstones, representing facies D1, is overlain by, and possible onlapped by conglomerates, white sandstones and flow tillite of facies D2 (Fig. 34E).

From here, walk south to southeast, to meet the boats at the pre-arranged time and place.

Day 4. Lower Smalfjord Formation and Cap Dolostones along North Varangerfjord and Leirpollen, Varanger Paleovalley (IV)

Introduction

During the first part of the day, the excursion examines outcrops on the north coast of Varangerfjord (Figs. 3 and 35A). The day starts with rocks showing the contact between the Smalfjord Formation and Vadsø Group on the northern flank of the Varanger paleovalley (Figs. 36 and 37). This includes deposits that show glacial and proglacial facies preserved along the steep unconformity with the underlying Vadsø Group. These exposures are important, since this is the only place where Arnaud (2008)

Figure 33. East Vieranjar'ga. All at Stop 3.4. All Smalfjord Formation except where stated otherwise. Note that A, C, and E form the top, middle, and base of a single profile, respectively. (A) Location 7. Thick low-angle cross-bedded sandy conglomerates of facies S1. (B) Location 8. Three sandstone fracture fillings in the Veinesbotn Formation; arrows point toward the fracture-filling margins. (C) Location 7. Matrix-supported conglomerates. (D) Location 9. Purple sandstone clast of the Veinesbotn Formation in facies D2. (E) Location 7. Base of Smalfjord Formation above Veinesbotn Formation. This includes a deformation diamictite and then matrix-supported conglomerates of Unit C1. (F) Location 9. Conglomerates overlying diamictites (all in facies D2). (G) Location 9. Well-stratified conglomerates showing clast imbrication and sandstones in facies S1.

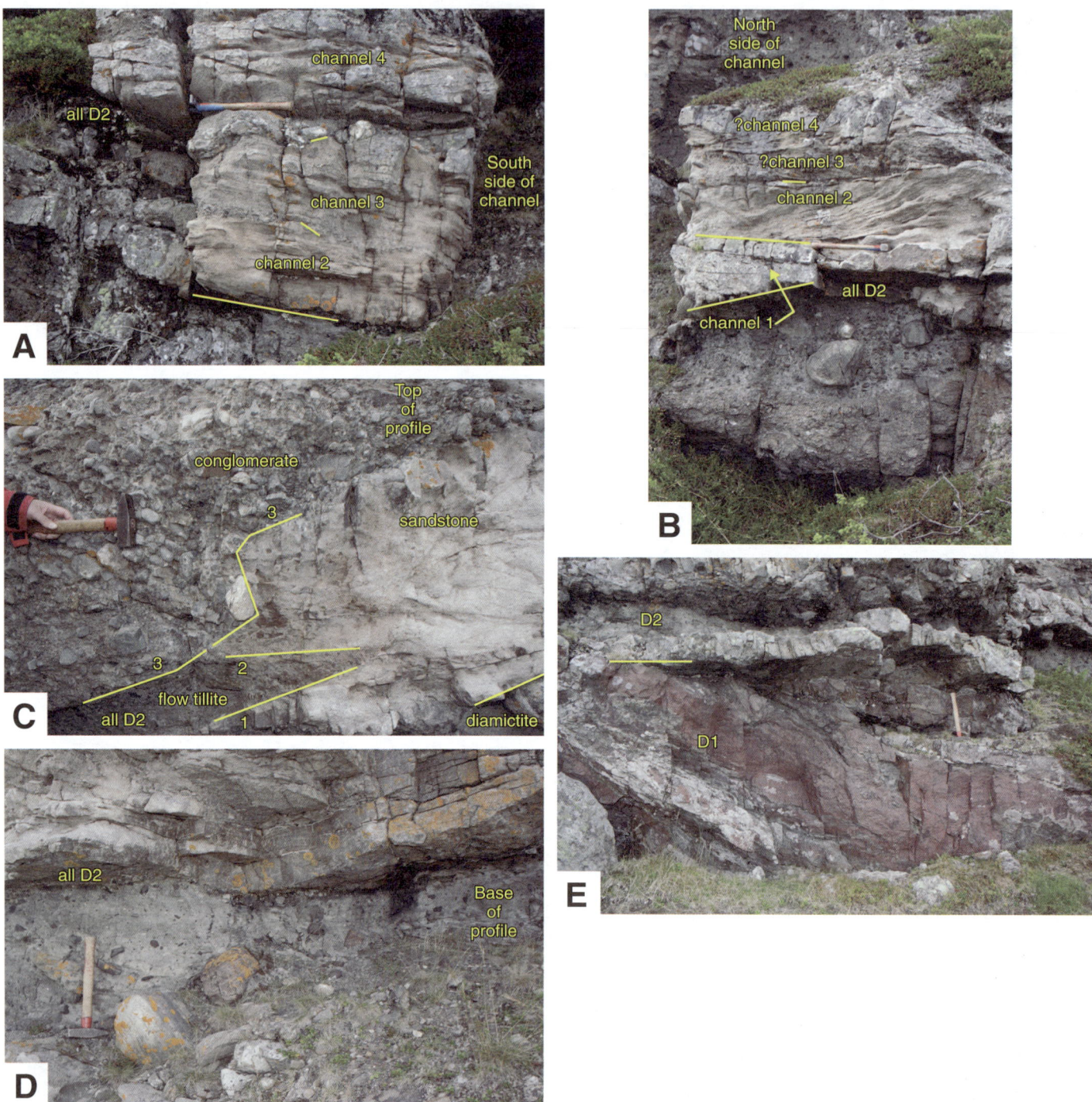

Figure 34. East Vieranjar'ga. All at Stop 3.4. All Smalfjord Formation except where stated otherwise. (A–B) Location 10. South and north sides of a complex, multiphase channel, all in facies D2. The first channel is incised into diamictites and is only seen on the north side. (C–D) Location 11. Upper and lower parts of a channel within facies D2. Large rounded basement and sandstone clasts occur within the diamictite, which is overlain by white-weathering sandstones, and this by conglomerates. Numbers 1–3 indicate major erosion surfaces; see text for details. (E) Location 12. Megaclast of Veinesbotn Formation within facies D1 of the Smalfjord Formation.

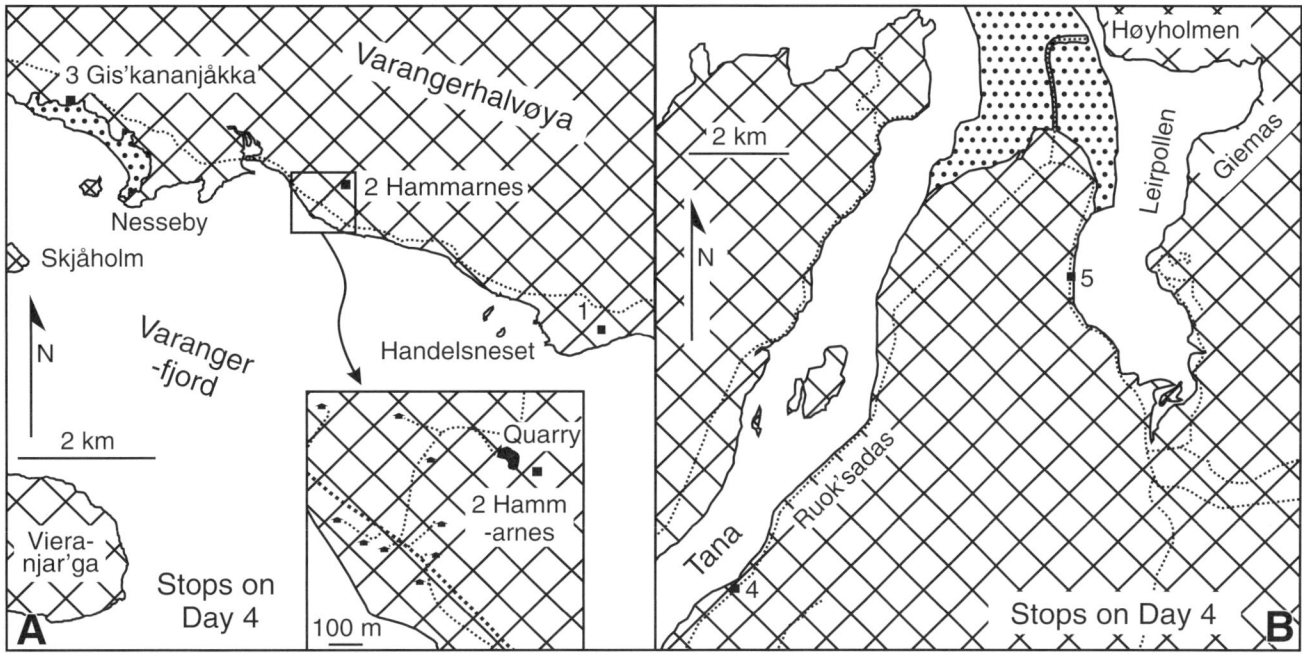

Figure 35. (A) Map showing the location of outcrops visited during Day 4 on the north side of Varangerfjord. Inset map shows location of Stop 2 in more detail. (B) Map showing the location of outcrops visited in the Tana River–Tanafjord area. Dashed lines indicate roads.

agrees with an interpretation of direct action by a glacier, as previously proposed by Edwards (1984, 2004). Most of the sediments exposed here were deposited subaqueously, in a proglacial or ice-contact environment.

In the latter part of the day, the guide documents outcrops showing the variation in facies in the dolomitic basal part of the Nyborg Formation (Member A), including outcrops in the Tana area (Fig. 35B). This typically varies between massive, buff-weathering dolomicrite with abundant sheet cracks through to sandstones and siltstones with an orangey-red color due to weathering of a minor dolomitic (?cement) component. Comparisons with other areas show marked similarities between parts of this

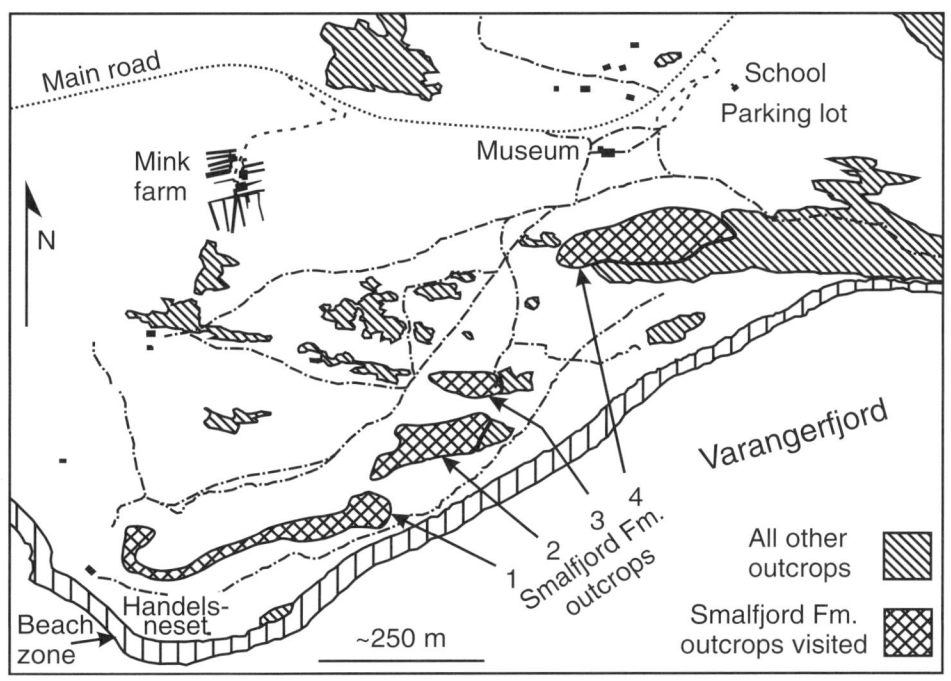

Figure 36. Stop 4.1. Map drawn from a satellite image showing the four outcrops visited at Handelsneset. Please be sure to stay on the paths shown when going to Location 1.

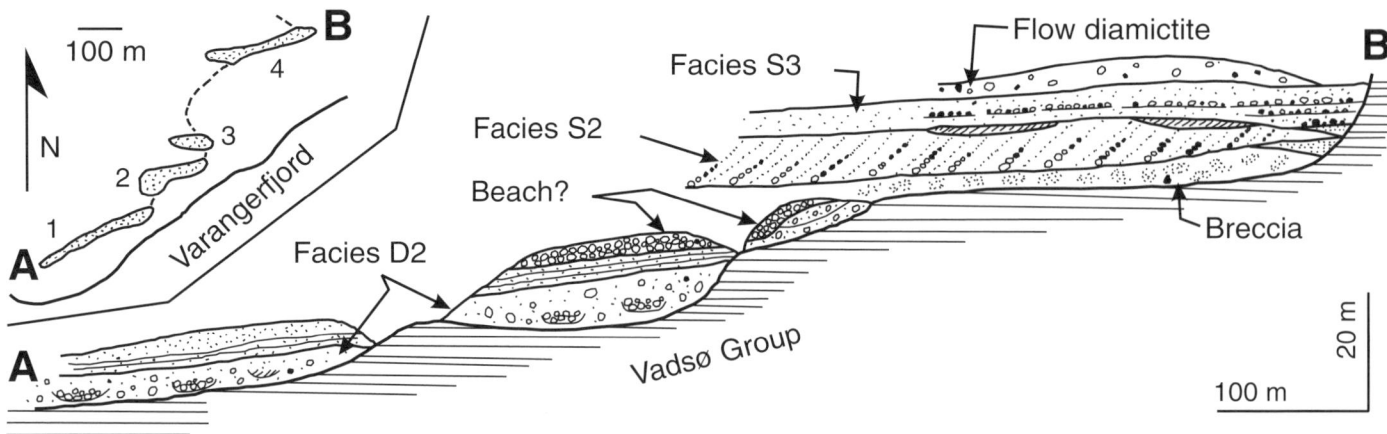

Figure 37. Stop 4.1. Semi-schematic profile through the four outcrops at Handelsneset, showing the sedimentary facies (after Edwards, 1984).

unit and typical Marinoan cap dolostones. This is the major constraint on the age of the Smalfjord Formation.

Stop 4.1: Handelsneset

Location. The outcrops are located in the vicinity of Handelsneset on the north coast of Varangerfjord; they also occur close to the Mortensnes Museum, which documents the Stone Age remains found on the raised beaches here. Park in the signposted museum parking lot and obtain permission in the museum to visit the outcrops, since you have to leave the official boardwalk. From the museum, follow the path down to the SW corner of Location 1 (Fig. 36) and then walk back obliquely uphill, visiting the outcrops in turn.

Note the prehistoric graves formed within the platy rock fragments as you walk on the boardwalk later.

Grid ref. GEC 70°07′42″N, 29°02′28″, the museum parking lot.

Introduction. In the course of the stop, four outcrops are visited, over a distance of several hundred meters. The traverse is best made from the bottom upward, admiring the ice-contact diamictites along outcrops one and two, possible beach conglomerates in outcrop three, and a variety of conglomerate and sandstone subaqueous facies in outcrop four (top; Figs. 36 and 37). If there is time, the participants are advised to visit the museum.

Description. Location 1. The exposures forming this outcrop have been well described by Arnaud (2008), who interpreted them as deposits of an ice-proximal, glaciofluvial to deltaic environment, occasionally subjected to overriding ice that caused a number of glaciotectonic structures. Edwards (1984) proposed an overall ice-contact fluvial and deltaic origin.

There are several important features to track as the exposure is followed. First, note the smooth upper surface of the Vadsø Group (Fig. 38A); striations occur on some of these surfaces (Laajoki, 2002). In some places, the unconformity also lies parallel to the bedding in the overlying Smalfjord Formation deposits. This raises the issue of how to determine an appropriate datum for inferring the depositional slope. Second, note the considerable range of textures and bedding styles in the Smalfjord Formation. Locally, there are large white granite boulders that suggest a winnowed remnant of a more substantial glacial deposit. Finally, be on the lookout for deformational as well as depositional textures; examples of low-angle shear structures are present (Fig. 38B). Arnaud (2008) interpreted these as evidence of an overriding glacier, although, in some cases, they may be preserved in blocks of frozen sediment that were either subglacially deformed, or deformed upon melt-out and redeposition (Fig. 38C).

Location 2. This location contains poorly sorted and weakly stratified sandstones and conglomerates similar to the previous location, but with a layer of conglomerate at the top. Toward the northeastern end, the contact with the Vadsø Group is very steep; in Figure 38D, the bedded white sandstones of the Vadsø Group dip gently to the right (east), while the unconformity cuts down steeply to the left (west).

Location 3. This exposure also has well-sorted and well-rounded conglomerates, composed mostly of dolomitic clasts from the Grasdal Formation, in the upper part of the Tanafjord Group, as seen in the previous location, although entrained blocks of sandstone also occur (Fig. 39A).

A distinctive, brown-weathering sedimentary breccia, consisting of mixed abundant intraformational clasts, rare extraclasts, and also concretions with silt nuclei, occurs as blocks up to at least 20 cm size within the conglomerates of the beach facies (Fig. 38E). These are exposed on the flat area above the cliff forming the south margin of the outcrop, only a few meters west of the path. These were derived from the fan or fan-delta deposits in the upper part of the Fugleberget Formation (Røe, 2003; Røe, 2008, personal commun.; Figs. 4 and 5). In some places here, the Smalfjord Formation is texturally mature enough to suggest beach deposits (Edwards, 1984).

Location 4. This is a complex exposure, with a wide range of textures and structures. The dominant feature are large foresets of conglomerates and sandstones with bedding dipping at ~10°W (Edwards, 1984). Parallel-stratified sandstones and poorly sorted conglomerates, some channeled into the stratified sandstones

Figure 38. Handelsneset. All at Stop 4.1. Locations and facies refer to Figure 37. (A) Location 1. Smooth, planar contact between Vadsø Group and buff-weathered channeled diamictites of facies D2 of the Smalfjord Formation at outcrop 1. Behind are white beach sandstones. (B) Location 1. Glacially sheared sediments forming folds (arrowed) in facies D2 of the Smalfjord Formation. (C) Location 1. Sheared and/or frozen clast with folds (arrowed) at the base of the Smalfjord Formation. (D) Location 2. Angular unconformity at base of the Smalfjord Formation (dashed line). (E) Location 3. Distinctive clast of the Fugleberget Formation (outlined by arrows) within the conglomerates of the Smalfjord Formation.

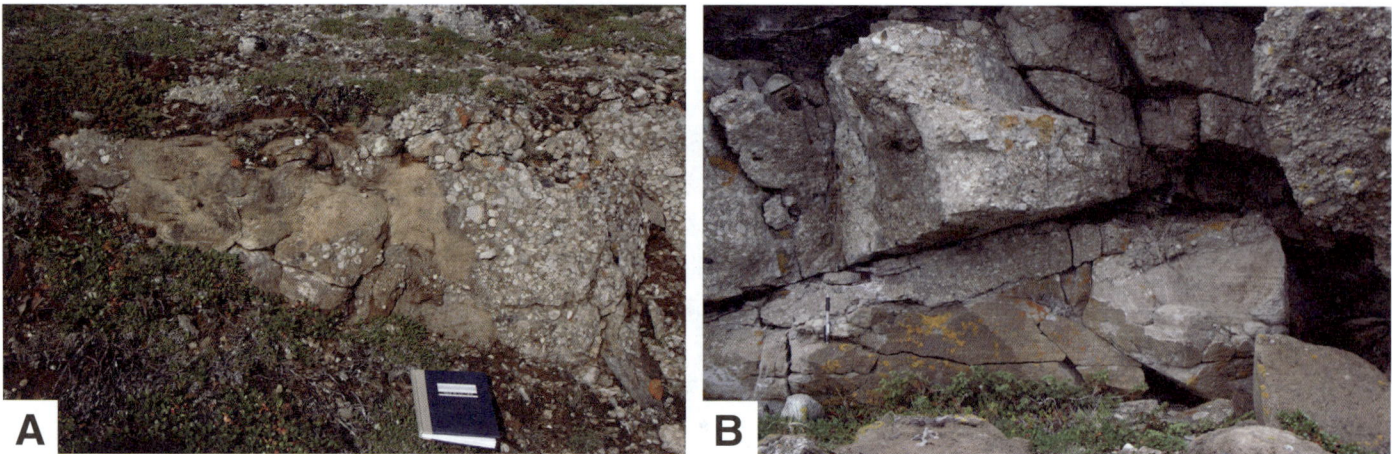

Figure 39. Handelsneset. All at Stop 4.1. Locations and facies refer to Figure 37. (A) Location 3. Conglomerate with sand clasts within the beach facies of the Smalfjord Formation. (B) Location 4. Channel in parallel-stratified sandstone filled with dolomitic clasts, Smalfjord Formation.

(Fig. 39B) and with abundant carbonate clasts, occur. Massive diamictites/conglomerates are also present toward the top.

Stop 4.2: Hammarnes Quarry

Location. When leaving the museum parking lot, turn left and drive west for ~4.5 km (Fig. 35A). Turn up the dirt road at GEC 70°08′31″N, 28°56′36″E toward a small quarry ~400 m from the road (Fig. 35A, inset). Park by the track leading down to the quarry. Walk southeastward through the southern margin of the quarry area and then eastward for 50 m, to the old sea cliffs, where the outcrops of dolomite begin. Investigation of the overhanging vertical rocks in the quarry is strongly discouraged.

Grid ref. NT 7365 8330, 1:50,000 map sheet Nesseby 2335 II, Edition 2-NOR. GPS 70°08′38.9″N, 28°56′18.1″E.

Introduction. Access to these outcrops is steep and, when wet, slippery. Be careful not to dislodge material that could fall on anyone below. For this reason, do not use hammers. Hard hats are recommended.

This location shows very rapid lateral and vertical changes in the facies of Member A of the Nyborg Formation (Fig. 7), the cap dolostone to the Smalfjord Formation. This rapidity is in part due to structural shortening in the area, although the kilometer-scale tectonics are not fully understood due to poor outcrop. In other areas, Edwards (1984) showed that usually there is a general increase in the clastic content of the cap dolostone upward, with massive dolomicrites at the base (when present) passing up to interbedded dolomicrites and red clastic sediments.

Description. The quarry was cut into gray sandstones and thin dark shales of the lower part of the Smalfjord Formation; due to south-directed shortening, these are now folded into a vertical orientation in the quarry, which is an extremely dangerous outcrop.

The old sea cliff, east of the quarry, shows three distinct profiles, from west to east: (A) massive sheet-cracked dolomicrite, (B) mixed massive, sheet-cracked dolomicrite, interbedded dolomicrite/red clastics, and white to dark-gray clastics, and (C) continuous interbedded thin dolomicrites and red clastic sediments.

Preliminary structural investigations suggest that each profile has been thrust over the next to the east. Thus, although the facies variations seem rapid, the sections/facies have been tectonically telescoped together. Sedimentary structures, including diagenetic doming in the sheet-cracked dolomites and graded bedding, and the bedding-cleavage relationships in the clastic sediments all show that the sequence is the right-way-up. Note that using sheet-crack doming to indicate way-up was considered very unreliable by participants with more expertise than the authors; here, at least, it is consistent with the other way-up criteria.

In profile (B; Fig. 40; GPS 70°08′38.9″N, 28°56′22.0″E), two fining-upward sequences have been proposed (Rice and Hofmann, 2001). Each passes from gray sandstones/conglomerates and dark shales through red clastics with thin dolomicrites (facies NA2 of Edwards, 1984) to massive dolomicrites (NA1). The development of facies NA2 under facies NA1 is atypical but not unknown in the Finnmark region (Edwards, 1984). No evidence was found by Rice and Hofmann (2001) of a tectonic contact at or near the boundary of the two sequences. From this, it was inferred that, because the gray sandstones and dark shales at the base of sequence 2 lay within the Nyborg Formation, the similar rocks at the base of sequence 1 most probably also lie within the Nyborg Formation, begging the question as to where the boundary between the Nyborg and Smalfjord Formations is located. In areas where diamictites form the top of the Smalfjord Formation and dolostones the base of the Nyborg Formation, this boundary is clear. Here, in contrast, massive carbonate deposition could only occur after clastic deposition had gradually ceased. This boundary is an unconformity, since the Varangerfjord outcrops of the Smalfjord Formation lie toward the base of the glacial succession (Føyn and Siedlecki, 1980; Edwards, 1984).

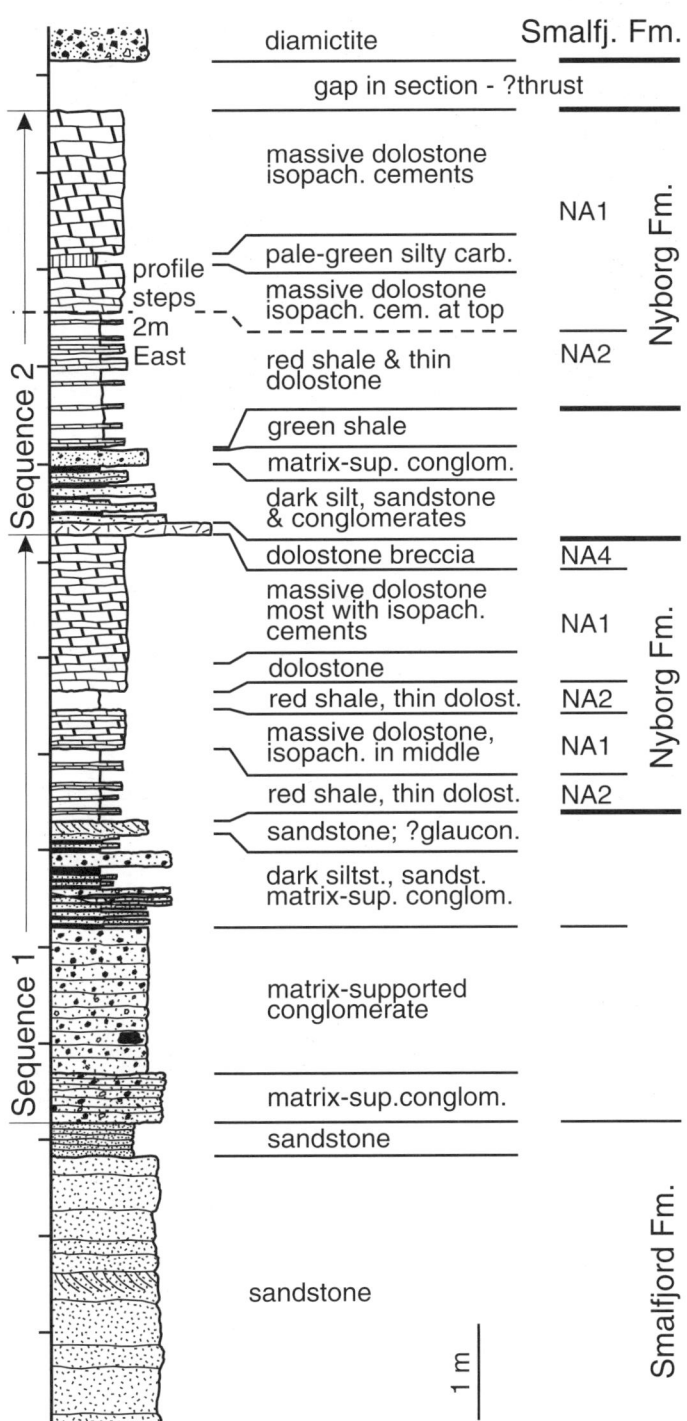

Figure 40. Log profile of the Hammarnes section (B) through the Smalfjord-Nyborg Formation boundary (after Rice and Hofmann, 2001) showing a large-scale cyclicity, with two postulated depositional sequences. Participants on the 2008 excursion preferred to place a thrust contact above facies NA4, close to the base of sequence 2, interpreting it as a tectonic repetition, not sedimentary cyclicity.

Note that although the massive dolomicrites have sheet cracks filled with isopachous cements, these are usually not developed near the contacts between the massive dolomicrites and other lithologies; this is typical for massive dolomicrites in the Finnmark region. In some cases, the lamination has been domed upward, but this is not stromatolitic in origin, since a brown (perhaps more ankeritic?) carbonate cement fills the core (Fig. 41A).

Twenty-nine samples from the profile all show essentially similar negative $\delta^{13}C$ values (Maieberg anomaly; Rice and Halverson, 2005–2010, personal observations).

Participants of the excursion in 2008 preferred to regard the succession as a thrust repetition of a single fining-upward sequence, implying that the clastic sediments at the base of each proposed cycle are part of the Smalfjord Formation. The thrust would lie above the thin layer of facies NA4, at the base of Sequence 2 in Figure 40. However, no evidence of thrusting, in particular slickensides or slickencrysts, has been found at the base of the "upper imbricate," despite the abundance of such structures on slip surfaces of all types, even with a minimal displacement, in the East Finnmark region. The Smalfjord Formation diamictites at the top of the section were emplaced by thrusting.

Stop 4.3: Gis'kananjåkka (East of Nesseby)

Location. Return to the main road and drive west for ~3.7 km (Fig. 35A). Park in the lay-by on the north side of the road (GPS 70°09′24.7″N, 28°50′20.8″E), just east of the small river Gis'kananjåkka. After asking permission from the house south of the road and immediately west of the bridge, walk down the path a few meters east of the stream, to where it opens out to the sea.

Grid ref. NT 6950 8446, 1:50,000 map sheet Nesseby 2335 II, Edition 2-NOR. GPS 70°09′21.4″N, 28°49′53.5″E.

Introduction. Facies typical of the basal part of the Nyborg Formation (Member A; Fig. 7) are well exposed; elsewhere, it can be seen that this is part of a Marinoan-type cap dolostone interbedded with red silts and fine sandstones (Edwards, 1984; Halverson et al., 2005). However, none of the typical characteristics of cap dolostones is present here, except negative $\delta^{13}C$ values typical of the Maieberg anomaly, not even sheet cracking with isopachous cementation. Other features present are typical (Member A) and atypical (Member B) edgewise breccias in the Nyborg Formation.

A clastic-dolomicrite interlayering is unusual for Marinoan cap-dolostone successions, which tend to be dominated by massive sheet-cracked dolostones (e.g., Halverson et al., 2005). In Finnmark, however, it is a common development, occurring at most localities along the north coast of Varangerfjord and also to the north at Ruok'sadas (Stop 4.4) and Grasdalen (Fig. 3), and westward onto Laksefjordvidda (cf. Edwards, 1984).

Description. The stream essentially runs down the contact between the Smalfjord Formation, seen in flat outcrops on the east bank of the river by the shoreline, and spectacularly folded

Figure 41. South and west Varangerhalvøya. All Day 4. (A) Stop 4.2. Cap dolostone with isopachous cements, domed-up and filled by brown carbonate, Nyborg Formation, facies NA1, Hammarnes. (B) Stop 4.3. Interbedded red shales and thin dolostones, Nyborg Formation, facies NA2, Gis'kananjåkka. (C) Stop 4.3. High-energy edgewise breccia of cap dolostone at the base of Member B, Nyborg Formation, Gis'kananjåkka. Imbrication indicates a top-to-left (west) flow direction. The large clast played a role in developing a steep topography. Later sediments onlapped and were draped over this clast. (D) Stop 4.5. NE-SW–trending Giemas antiform, Leirpollen. (E) Stop 4.5. Cap dolostone with buckled and brecciated isopachous cements, possibly overprinting stromatolites, Member A, facies NA1, Nyborg Formation. Dashed lines are the axial surfaces of a box fold. Solid lines with half-arrows are minor detachments developed between folded and unfolded areas. Leirpollen. (F) Stop 4.5. Close-up of E, showing later brown carbonate cements that filled voids in brecciated sheet-cracked dolostone (solid arrows, at base) and platy fragments (open arrows, at top).

interbedded dolomicrites and red clastic sediments of Member A (facies NA2) of the Nyborg Formation (Fig. 41B). The contact between the two formations is not exposed.

Fold axes trend WNW-ESE and are upright to south vergent (and upward facing), with NNE-SSW–oriented bedding-plane slickensides indicative of flexural slip. Fold wavelengths are shorter because the bedding is thinner here. At low tide, it can be seen that these folds continue out to sea. Fold accommodation structures are present in some hinge zones, as minor rootless thrust faults and hinge collapse structures.

Although most of the succession is composed of dolomicrites and red clastic sediments, occasional channels filled with an edgewise breccia ("mud-flake" conglomerate; facies NA4; Edwards, 1984) can be seen, attesting to intermittent syndepositional higher-energy events scouring the interbedded dolomicrites and red clastic sediments.

If the shoreline outcrop is followed to the west (GEC 70°09′25″N, 28°48′42″E), the top of Member A is reached, marked by spectacular examples of edgewise breccias at the base of Member B (Fig. 41C). In these, massive and crudely horizontally stratified sandstones and conglomerates, with dolomicrite clasts 10 cm or more across and up to 3 cm thick (but mostly much smaller), were deposited in strongly erosive scours 30 cm deep, together with spherical subrounded to subangular quartz pebbles up to 0.5 cm across. In other cases, the breccia was deposited as discrete flakes in laterally persistent layers within the sandstones. The breccia thickness and clast size show rapid lateral variations. This edgewise breccia passes up to thick-bedded, dark, reddish, relatively coarse-grained sandstones and conglomerates of Member B.

Nine samples collected from the edgewise breccias of Member B and a few meters down into the folded interbedded succession of Member A all showed negative $\delta^{13}C$ values (–5.72‰ to –3.60‰ VPDB; Rice and Halverson, 2005–2010, personal observations), with no systematic variation from bottom to top.

Stop 4.4: View to Ruok'sadas

Location. Drive west, through Varangerbotn to Tana Bru and thence further north for ~26 km, along the east side of the Tana River (Fig. 35B).

Grid ref. NU 4725 1237, 1:50,000 map sheet Tana 2335 IV, Edition 2-NOR. GPS 70°24′42.1″N, 28°15′46.0″E.

Introduction. Deformation dies out toward the east within the Gaissa Thrust Belt, with only a few thrusts developed on and east of Laksefjordvidda (Fig. 3). Immediately to the east of the Tana River, one such large-scale thrust crops out, below a complex, partly collapsed, thrust-related fold on Ruok'sadas, in rocks at the boundary of the Tanafjord and Vestertana Groups.

Description. The hillside directly ahead shows a prominent dark red scree, which is orange on the west side. This is a west-dipping, inverted sequence of the Nyborg Formation, with interbedded thin dolomicrites and red shales (Member A; facies NA2) at the stratigraphic base (now physically the top). No massive cap dolostone crops out at this locality. Seventeen samples from this sequence gave consistently negative $\delta^{13}C$ values (Rice and Halverson, 2005–2010, personal observations). These are overlain by inverted orange-weathered diamictites of the Smalfjord Formation and thence by the Gamasfjell Formation (Tanafjord Group). Somewhat to the southeast, large-scale folded blue-green shales of the Stappogiedde Formation (Innerelv Member), also forming an extensive scree slope, underlie this sequence, in some places along a minor thrust, probably linked to the nearby thrust to the west. Climbing up to the outcrop is hard work and unrewarding, except for the view, especially of the sandbars in the Tana River.

Stop 4.5: Leirpollen

Location. Continue to drive north. After ~10 km, the road turns toward the southeast just beyond the indicated turning to Høyholmen, to the north. Some 3 km after this, look for several ~1-m-sized blocks of buff-weathered rock set in a grassy bank on the right (southwestern) side of the road, which is here heading southeastward (Fig. 35B).

Please do not use hammers at this outcrop, which is unique in the Finnmark region.

Grid ref. NU 5385 1890, 1:50,000 map sheet Tana 2335 IV, Edition 2-NOR. GPS 70°28′10.9″N, 28°26′38.6″E.

Introduction. This outcrop shows a very condensed sequence of the Nyborg Formation, being in all only 8 m thick (Edwards, 1984). Although this might be partly due to erosion at the base of the unconformably overlying Mortensnes Formation, the thinness here of Member A (usually quoted at ~1 m, but actually considerably more) compared to other areas suggests it may also be primary in part. The concentric structures in much of the rock are enigmatic: Are they stromatolites or soft-sediment deformation structures or a combination of both?

Note that the large-scale Giemas antiform in the Tanafjord Group, exposed on the east side of the fjord (Leirpollen), is a symmetrical structure when viewed along its fold axis (Fig. 41D).

Description. The lower part of the Nyborg Formation here consists of ~1 m of micritic dolostone (facies NA1; Fig. 41E). This can be divided into four parts, one of which (B) displays a marked domal structure, although there are intervals within this where doming is minimal. This is overlain and underlain by essentially flat-lying dolomicrites (part A is not visible in Fig. 41E). Sheet cracks, which are common in parts B and C and are parallel to the layering even when bent around the domes, are filled by translucent isopachous cements up to a couple of millimeters thick that are essentially constant in thickness around the domes.

The lower (not seen in Fig. 41E) and upper parts (A and D) of the dolostone were not affected by sheet cracking and secondary cements, as also observed at Hammarnes (Stop 4.3), and show no signs of doming.

Whether the domal structures are stromatolitic in origin, as has generally been inferred (Siedlecka and Roberts, 1992), or are more-or-less postsedimentary, possibly diagenetically induced, structures (Halverson and Maloof, 1999, personal commun.) is not certain, but the weight of evidence lies strongly toward the latter option. Little or no sign of thinning of the beds on the sides of the domes, typical of stromatolites, is apparent. In contrast, layers can be traced from one dome to another with no noticeable change in thickness. There is no obvious (or not obvious, either) "interstromatolite" sediment, which frequently appears lithologically different, especially when weathered, in unequivocally stromatolitic deposits (cf. Grey et al., 2011). In some areas, especially under layers that are unfolded, the "interstromatolite" area is filled with a breccia including platy fragments, often with a very high aspect ratio, that are typically planar (Fig. 41F, arrowed), wholly inconsistent with the curved and laterally thinning geometry of a stromatolite. These platy fragments indicate that the dolomicrites had enough strength at this stage to undergo brittle failure, although elsewhere they were still soft enough to be folded plastically.

In some cases, the doming of the dolomicrites can unequivocally be ascribed to buckling and brittle fracturing/thrusting of the bedding. Deformation-induced voids, from only 0.5 mm up to ~3 cm thick, some containing fragments of the sheet-cracked sediments, were filled with a darker-brown, perhaps ankeritic, cement (Fig. 41F). These filled voids are often dome shaped, both concave upward and downward.

The structures have the geometry of class 1B parallel folds, keeping a constant bed thickness from hinge zone to limb. The voids in the hinge zones are suggestive of flexural-slip deformation. In some cases, box folds formed, with diverging axial surfaces (Fig. 41E). Elsewhere, it seems that in the absence of a vertical load, extension (brecciation) on the outer and shortening (thrusting) on the inner parts of folds imply a component of tangential longitudinal strain. This is seen as a tendency for brecciation to have occurred at the top of part B, before part C was deposited. If a purely soft-sediment-diagenetic origin to the doming is assumed, line-length shortening would be ~30%, the source of which remains uncertain.

The large hinge zones and curved limbs of the folds suggest that the rocks had a low mechanical anisotropy (a high anisotropy gives chevron folds). On the reasonable assumption that the growth of isopachous cements would have imparted a marked anisotropy, this suggests that sheet cracking occurred after folding. Counter to this, the isopachous cements appear to have been offset by fracturing in some places. If the structures are diagenetic in origin, then slip must have occurred between the flat-lying part A and the folded part B (lift-off folds). Similar slip must also have occurred on both the upper and lower surfaces of unfolded layers within part B (Fig. 41E).

The lack of 3-D outcrop makes it more difficult to unequivocally differentiate between the two possibilities (biogenic versus soft-sediment deformation/diagenetic), but at the very least, there was a major component of pre- to syndiagenetic folding.

The carbonates rest on ~1 m of poorly exposed mudstone that, in turn, rests on a massive dark rock with only a few small carbonate clasts. This may be a loessite, comparable to that found at Grasdal to the north (Edwards, 1979).

Some 40 m to the northwest of the massive dolomicrites, in outcrops set a few meters to the southwest of the road, two thin (<7 cm thick) dolomicrite layers occur within the red shale/sandstone sequence, comparable to Facies NA2. Since this implies that the whole succession down to and including the massive dolostones are a part of Member A, the initial assumption that the Nyborg Formation is stratigraphically thin here must be reassessed.

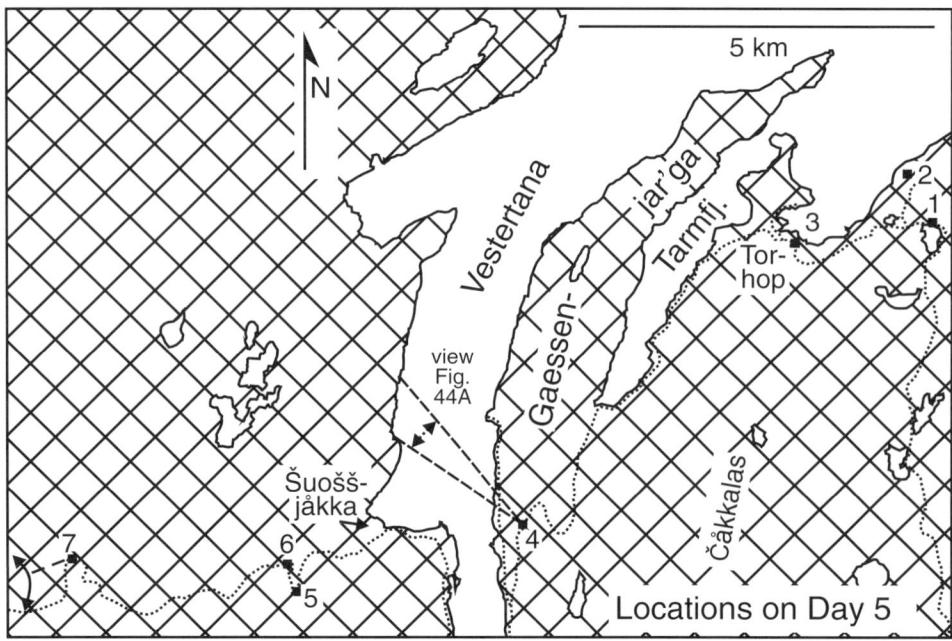

Figure 42. Map showing the location of outcrops visited during Day 5. Dashed line indicates roads.

Day 5. Glacigenic Rocks in the South Tanafjord–Vestertana Area

Introduction

The outcrops on this day (Fig. 42) cover a number of sedimentological units not seen elsewhere during the excursion. In part, they are included as an alternative day if the weather is too bad to make some of the boat trips required on other days. The day includes a diverse range of outcrops of the Dakkovarre Formation (Tanafjord Group), Smalfjord, Nyborg, and Mortensnes Formations and Lillevatn Member of the Stappogiedde Formation. Units seen here that are not seen elsewhere during the excursion are those at Stops 5.3, 5.6, and 5.7.

Stop 5.1: Šœresgied'djav'ri

Location. Drive north from Tana Bru, on the west side of the Tana River. Turn left at Ruos'tefiel'bma, just before the last gas station in the immediate area, and continue westward, past the road turning north to Smalfjord village. Drive around Smalfjord and then for a further ~4.3 km from where the road starts to climb uphill on its west side. Just after the road goes downhill steeply beside rust-stained white sandstone outcrops, it turns sharply west, around the north end of a small lake (Fig. 42). Park by this, opposite the orange-colored diamictite outcrop.

Grid ref. NU 3865 2002, 1:50,000 map sheet Smalfjord 2236 I, Edition 3-NOR. GPS 70°28′56.7″N, 28°02′04.1″E.

Introduction. The subglacial unconformity under the Smalfjord Formation cuts down section through the Tanafjord Group from north to south; in the south (Vestertana–Torhop–Smalfjord–Ruos'tefiel'bma area), it cuts through the base of the Gamasfjell Formation, into the Dakkovarre Formation (Fig. 4).

Description. At the south end of Smalfjord, interbanded quartzites and dark siltstones and shales of the Dakkovarre Formation underlie the Smalfjord Formation along the roadside. Here, 6 km north, quartzites of the Gamasfjell Formation (on the east side of the outcrop), dipping moderately steeply to the west, are overlain by buff-weathering (dolomitic) diamictite of the Smalfjord Formation (Member b; Figs. 8 and 43A) with an anastamosing spaced cleavage dipping more steeply to the west than the Gamasfjell Formation bedding. Clasts of oolite, micrite,

Figure 43. South Tanafjord. (A) Stop 5.1. Typical dolomitic diamictite of Member b of the Smalfjord Formation. Location of clast shown in B is marked by yellow square, Šœresgied'djav'ri. (B) Stop 5.1. Clast of diamictite within sandstone, Šœresgied'djav'ri. (C) Stop 5.3. View of klippen of inverted Gamasfjell Formation (arrowed) across Indre Torhop. (D) Stop 5.3. View of Tanafjord from near Torhop.

and stromatolite, derived from the Grasdal Formation are present. A lenticular bed of sandstone, up to 40 cm thick, locally parallel-laminated and containing a few clasts of tillite (probably frozen solid during transportation), was most likely deposited by subglacial meltwater (Figs. 43A and 43B; Edwards, 1984). As sediment accumulated in the subglacial channel, it filled up and became closed-off.

Toward the western end of the outcrop, the member becomes more pelitic and hence has a better-developed Caledonian cleavage.

Stop 5.2: Auskarnes

Location. Drive west for ~300 m from Stop 5.1 and turn right onto the dirt road, passing by the buildings with the attractive variegated slate roofs. Drive north for 0.9 km and park opposite the obvious small white hill on the west (cf. Fig. 43C, on the right side of the fjord, photo taken from west of Stop 5.3).

Grid ref. NU 3835 2068, 1:50,000 map sheet Smalfjord 2236 I, Edition 3-NOR. GPS 70°29′19.0″N, 28°01′43.1″E.

Introduction. In the area south-southeast of Laksefjord (Fig. 2), the regularly spaced thrusts, at every ~0.5–1 km, mapped within the western part of the Gaissa Thrust Belt die out (Townsend et al., 1986, 1989), and the shortening drops from ~50% to only 15% (Chapman et al., 1985). Within this easterly, low-strain zone, there are very few thrusts (Fig. 3). One is exposed to the south of Torhop and extends into this area, where a structurally inverted and strongly sheared succession at the contact of the Tanafjord and Vestertana Group can be seen. There are more thrusts on the Digermul Peninsula and also south of Trollfjorden, with NE-SW orientations; these latter link with N-S–oriented thrusts immediately east of the Tana River (Fig. 3).

Description. The small white hill is composed of quartzitic sandstones of the Gamasfjell Formation; cross-bedding shows that this sequence is inverted. The base of the quartzites shows a concave-upward shape as a series of lateral ramps cutting down first through a thin band of gray diamictite of the Smalfjord Formation and then thin interbedded dolomicrites and red shales, and then red and green shales/sandstones (Members A and B, Nyborg Formation; Fig. 7). The inverted rocks (bedding is steeper than cleavage) form the upper limb of an overturned E-verging upward-facing syncline. The fine lamination in the Nyborg Formation and the paucity of clasts in the diamictite indicate high strains within these sediments, although close examination reveals abundant small ?comminuted clasts. The thrust at the contact between the inverted Gamasfjell and Smalfjord Formations may be the main thrust, juxtaposing inverted hanging-wall–anticline Gamasfjell Formation against inverted footwall-syncline Smalfjord Formation, or it might be a relatively minor thrust, through a collapsed footwall-syncline. The trace of the synformal axial surface lies in the vegetated area to the east, beyond which the Nyborg Formation is the right-way-up and at a relatively low level of strain.

Looking 4–5 km to the SSW, irregular small hills can be seen on top of the larger-scale rounded hill (Čåkkalas, Fig. 42; Grid Ref 1:50,000 Smalfjord 2235 I NU 3600 1650). These are klippen of the Dakkovarre Formation, tectonically overlying interbedded dolomicrites and red shales at the base of the Nyborg Formation (Fig. 3). This is part of the same thrust system as seen at the outcrop here.

Stop 5.3: Torhop

Location. Drive back to the main road and turn right. Continue downhill and around the bay with the scattered farm buildings/houses of Torhop village. Stop on the corner of the road shortly after turning sharpish to the left, with the harbor in Indre Torhop and Tanafjord in view, within outcrops having an obvious dark-purple hue (Fig. 42).

Grid ref. NU 3670 1973, 1:50,000 map sheet Smalfjord 2236 I, Edition 3-NOR. GEC 70°28′49″N, 27°58′58″E.

Introduction. The Dakkovarre Formation is a 275–350-m-thick mixed sandstone-shale sequence in the middle part of the Tanafjord Group. In the upper part of the formation, there is an up to 130-m-thick Ferruginous sandstone member (Fig. 4; Siedlecka and Siedlecki, 1971), a distinctive dark-purple massive sandstone, often looking like burnished rusty iron. The locality also provides a superb view of the Late Neoproterozoic sediments along Tanafjord.

Description. At this outcrop, purple ferruginous sandstones of the upper part of the Dakkovarre Formation crop out in massive beds up to ~1 m thick. Clasts of this distinctive unit are described from within Smalfjord Formation diamictites of Member a later in the excursion (Stop 6.2, Location 1).

In fine weather, the view northward along Tanafjord shows much of the stratigraphy of the Gaissa Basin (Fig. 43D). On the east side, white cliffs reflect the abundance of sandstones/quartzites within the Tanafjord Group, folded on a large scale; these are likely buttressing folds against a postulated buried step in the basement architecture. On the west side, relatively smooth vegetated ground, rising quite steeply from the sea, is made of the Vestertana Group on the Digermul Peninsula (Fig. 2); these are visited on Day 7. The Vestertana Group is overlain by the steeper-sloped rounded hills of the Digermul Group, which reaches up into Tremadoc black shales (Reading, 1965; Fig. 5). Medium- to large-scale folding is also present in these rocks, mostly on the other side of the peninsula. As the coast is parallel to the fold axes, the folding is not apparent.

Stop 5.4: SW Duoivejeakkečohkat (East Side Vestertana)

Location. Continue driving westward, past the turning down to Torhop, around the sharp turn south and down the east side of Tarmfjord. Drive up the hill at the southern end of the fjord and continue through the double hairpin bend until you can see down into Vestertana. Stop at the obvious large parking place ~100 m beyond the second hairpin bend, on the west side of the road (Fig. 42).

Grid ref. NU 3310 1590, 1:50,000 map sheet Smalfjord 2235 I, Edition 3-NOR. GEC 70°26′45″N, 27°53′03″E.

Introduction. The view to the west, across Vestertana, shows the contact between the Nyborg and Mortensnes Formations. This is complex on a regional scale, since the latter gradually cuts down-section through the former toward the south; only Members A–C of the Nyborg Formation (Fig. 7) are present around Vestertana. In the same area, the Middle Member of the Mortensnes Formation cuts out the Lower Member, while itself thickening toward the north (Fig. 9). The contact between the Mortensnes Formation and the overlying Lillevatn Member (Stappogiedde Formation) is also exposed.

On the east side of the road, a very large clast of basement rocks occurs within the Smalfjord Formation.

Description. The lower part of the cliff (Fig. 44A), showing subhorizontal apparent bedding (the rocks are actually strongly folded, but the cliff is subparallel to the fold axes), constitutes the Nyborg Formation. Member A (dolomite bearing) is not visible in the profile, but both Members B and C are present, although the boundary is hard to see (and hence is not shown in Fig. 44A). These are overlain by a thin (~1.5 m; Edwards, 1984) development of the Lower Member of the Mortensnes Formation, at the base of the obvious buff-weathering (dolomite bearing) cliff, which is formed of the Middle Member (Thick Submember; Fig. 9). The boundary between the Middle Member and gray-colored Upper Member of the Mortensnes Formation is best seen toward the northern end of the panorama. In contrast, the overlying boundaries, defining the extent of the darker, finer-grained Lower Submember and the lighter-colored, coarser Upper Submember of

Figure 44. South Tanafjord. (A) Stop 5.4. Panorama view of the west side of Vestertana showing the stratigraphy of the Vestertana Group from near the base of the Nyborg Formation (Member B) up to the Innerelv Member of the Stappogiedde Formation (Fig. 4). Field of view is shown in Figure 42. SW Duoivejeakkečohkat. Abbreviations: MB—Member B, MC—Member C, LM—Lower Member, MM (TSM)—Middle Member (Thick Submember), UM—Upper Member, LM (LSM)—Lillevatn Member (Lower Submember), LM (USM)—Lillevatn Member (Upper Submember), I (red)—Innerelv Member (red shales), I (blue-green)—Innerelv Member (blue-green shales and siltstones). (B) Stop 5.4. Large crystalline basement clast within Member d of the Smalfjord Formation. Note the laminations dipping ~30° to the west (right) in the background (marked by line). SW Duoivejeakkečohkat. Picture courtesy of Tony Spencer. (C) Stop 5.6. Green shales of Member C of the Nyborg Formation underlying the Lower Member of the Mortensnes Formation. Lift-off folding is seen along a thrust parallel to the Nyborg-Mortensnes Formation boundary (partially under the vegetation; open arrows) but within the Nyborg Formation, resulting in an accommodation thrust that is roughly horizontally oriented (solid arrows). The uppermost part of the Nyborg Formation is not folded; the contact with the deformation diamictite (Lower Member) of the Mortensnes Formation is indicated. East of Poas'tagurra.

the Lillevatn Member of the Stappogiedde Formation are best seen toward the southern end. Above this, but not seen during the excursion, the basal part of the overlying Innerelv Member is pelitic and bright red, while the stratigraphically higher parts are blue-green and more quartz rich.

On the east side of the road, toward the southern end of the outcrops lying opposite the parking place, an exceptionally large lonestone of crystalline basement gneiss is present within green-colored shale rocks of Member c of the Smalfjord Formation (Fig. 44B). Another large lonestone lies above. Toward the south, laminations in the formation can be seen to dip at a lower angle than the cleavage.

Stop 5.5: Šuoššjåkka

Location. From the parking place, drive south, downhill into, and then westward around the southern end of Vestertana. Continue past the turning to Sjursjok (Šuoššjåkka) village. Some 1.9 km beyond that turning, there is an obvious uphill 180° extreme hairpin bend as the road crosses the Šuoššjåkka (river; Fig. 42). Park in this area, well off the road, and go to the outcrops on the eastern, downhill side of the river, essentially at the corner.

Grid ref. NU 2981 1460, 1:50,000 map sheet Smalfjord 2235 I, Edition 3-NOR. GPS 70°26′08.2″N, 27°47′47.1″E.

Introduction. The Lillevatn Member (40–110 m; Stappogiedde Formation; Figs. 4, 5, 6, and 7), which overlies the glacial sediments of the Mortensnes Formation along a sharp or rapid transition, includes Upper and Lower Submembers, both of which thin from south to north. The Lower Submember, seen here, varies from 3 to 55 m thick (Edwards, 1984).

Description. The outcrop on the east side of the corner is composed of gray-colored to buff-greenish–colored, thin- and parallel-laminated mudstones and siltstones (Lower Submember of the Lillevatn Member). Some 10 m northward along the road, a *very* similar finely laminated rock crops out, but this has rare lonestones of dolostone and basement. This is the uppermost part of the Mortensnes Formation. In the field, it is extremely hard to actually define the boundary between the two lithologies here. In thin section, however, it can be seen that the Lillevatn Member is well sorted, while the Mortensnes Formation is not. Directly uphill, to the east, the Upper Submember of the Lillevatn Member crops out.

Stop 5.6: East of Poas'tagurra

Location. From the hairpin bend, drive uphill until the road goes round a sharp left-hand bend (Fig. 42). Park where safe beyond the bend and go to the obvious Nyborg-Mortensnes Formation boundary.

Grid ref. NU 2968 1505, 1:50,000 map sheet Smalfjord 2235 I, Edition 3-NOR. GPS 70°26′21.9″N, 27°47′38.7″E.

Introduction. Due to erosion at the base of the Mortensnes Formation, generally only Members A and B of the Nyborg Formation (Fig. 7) are exposed at easily accessible sites. Member C, however, is exposed here, at the southern end of Vestertana, under the Lower Member of the Mortensnes Formation.

Description. Turbiditic graded green mudstones and shales in rhythmic beds of Member C of the Nyborg Formation have been folded about N-S axes, under the Mortensnes Formation. Due to their different effective bed thicknesses, folds of different wavelengths have formed, with intermediate- to regional-scale structures in the Mortensnes Formation and outcrop-scale structures in the Nyborg Formation (Fig. 44C). Although the folds in the thin-bedded Nyborg Formation die out toward the Mortensnes Formation contact, the shortening has resulted in some flexural-slip displacement (lift-off folds) along a contact that lies within the uppermost part of Member C.

The Lower Member of the Mortensnes Formation, directly above the Nyborg Formation, mostly consists of locally derived reworked green mudstones/shales of Member C; no basement clasts are present. However, a few meters to the east, immediately across an obvious ~30-cm-wide fault breccia, ~3 m of diamictites of the upper part of the Lower Member, dipping steeply eastward, contain some basement clasts. This is overlain by faintly buff-colored diamictites of the Thin Submember of the Middle Member of the Mortensnes Formation. Further downhill, nearer the corner, a very thick sequence of the Upper Member is exposed, with basement clasts up to 0.5 m size. Many of these have been fractured essentially normal to the Caledonian cleavage, which is wrapped around the clasts. In some cases, especially on fresh surfaces, a Caledonian tectonic strain shadow can be discerned in the diamictite, adjacent to the clasts. Within the strain shadow, no cleavage is present, but cracks orthogonal to the external cleavage, and generally slightly curved, attest to brittle deformation in the matrix rock "supported" by the clast. This confirms that they are Caledonian tectonic structures. In contrast to the "strain shadows" described at Stop 6.4, Location 1, which are of sedimentary (glaciotectonic) origin, the lithology at this stop is the same inside and outside the strain shadow.

Stop 5.7: Geaidnonjoasèohkat

Location. From the last outcrop, continue driving westward. Initially, the road goes uphill, then down into a broad wooded valley, and then begins to climb uphill again. Ahead, there is the steep east-facing hillside of Geaidnonjoasèohkat; the rocks of this stop lie on this hillside. As the road rises above the tree line, park in the large gravelly area on the inner side of a long left-hand uphill bend (Fig. 42; GEC 70°26′23″N, 27°42′44″E). From the car-park, cross over the road and walk up the hillside close to the reindeer fence immediately to the northeast, until you reach the obvious steep outcrop of buff brown diamictites. On the way, note the marked change in vegetation across the fence; to the north, it has been extensively browsed and trampled by the reindeer, which tend to migrate south toward the fence in late summertime.

Grid ref. NU 3310 1585, 1:50,000 map sheet Smalfjord 2235 I, Edition 3-NOR. GEC 70°26′17″N, 27°41′42″E.

Introduction. The Middle Member of the Mortensnes Formation consists of a Thick Submember, seen to the north (see Day 7), and a Thin Submember, seen here. This stop is essentially column 3 in Figure 9, which shows that the boundary between the two submembers lies only a short distance to the north. Although the two submembers differ in thickness and deposits, both are dominated by carbonate-bearing (matrix and clasts) diamictites and associated rocks, although there is typically a very large siliciclastic component to the matrix as well. The carbonate was almost certainly derived from dolostones of the Grasdalen Formation of the Tanafjord Group. The Middle Member here is ~3 m thick and forms a prominent cliff in the hillslope, with a gradational contact to the gray-green rocks of the underlying Lower Member. The Thin Submember of the Middle Member was deposited subaqueously, with several process contributing toward the sediments observed, including (1) sediment rain-out from icebergs, (2) iceberg turbation, and (3) iceberg dump structures. The latter form when an iceberg tilts and sheds a large amount of sediment at one time.

Description. Detailed descriptions of a few features are given with GPS points, but similar structures can be found throughout the submember. Although the descriptions given mostly refer to outcrops north of the reindeer fence, the section can be followed for ~400 m to the south, and the road can then be used to walk back to the vehicles.

Cross the fence and walk north, beside the Middle Member, to GPS 70°26′20.5″N, 27°41′38.8″E. Here an ~20-cm-thick distinctively pale-yellow to buff-colored band of dolostone occurs, with angular to subrounded clasts of carbonate up to several centimeters in size (Fig. 45A). This comes in as a single unit; similar units have 100% basement-derived clasts or just sandstone clasts. Such bodies are very discontinuous laterally, forming pods that are, in some instances, only a few centimeters thick and that cannot be traced laterally. The fact that the overlying sediment onlaps the margins of the pods attests to a depositional topography and hence strength within these deposits (Fig. 45B); the matrix has a higher carbonate-mud content than that of the adjacent rocks. Although the origin of such deposits is unclear, we speculate that they may have formed when a block of sediment-laden ice sank under its own weight and melted, depositing its contents in situ (melt-out diamcitite). The example seen is overlain by the upper part of the Middle Member, which gets finer-grained upward.

Somewhat further downhill, following the outcrop, a small- to medium-sized loose block of the Middle Member (possibly from close to the Lower Member) contains an obvious fold (Fig. 45C) cut by a prominent cleavage at a high-angle to the axial surface. Since the cleavage is Caledonian in age, the folding must be of soft-sediment (likely glaciotectonic) origin. Note that the well-developed bedding on the south side of the block is cut out by massive diamictite on the north side (directly under the pencil). There also appears to be a deformation diamictite underneath the south (left) end of the pencil; the pronounced layering in the sediments disappears a few centimeters below the contact with the diamictite (arrowed).

Complex sedimentary structures, including folds and possible dump structures are relatively common in the Thin Submember. Figure 45D, from between the two previous descriptions, shows the typical character of the diamictite in this submember. Above the pencil, well-stratified diamictite shows alternations of structureless and laminated diamictite with a pronounced lenticularity. This sometimes looks like tectonic pinch-and-swell structures (arrowed), which would indicate lamination-parallel extension. Below the pencil, there is an example of a lens of structureless diamictite that pinches out very quickly. This may be an erosional remnant; the laminations above are lenticular and dip quite steeply northward. These show small, slightly asymmetric (top-to-north) folds, indicating slight downdip slumping.

Figure 45E, at GPS 70°26′16.8″N, 27°41′43.7″E, south of the fence, shows an example of downlapping inclined laminations laterally accreting to the right (northwest). The laminae thicken and thin as they pass from the top to bottom. This is not inconsistent with a dump structure; as material cascaded down from an iceberg and hit the sea bottom, it spread out laterally. Thus, in such a model, the structures are not caused by a current reworking the sediment, but by the current induced by the falling sediments themselves; essentially, it is a gravity flow.

Figure 45F, from ~50 m south of the fence, shows very peculiar laminations associated with the large white clast, which might be a dropstone. The laminations immediately to the right of the clast are downlapping eastward onto the prominent layer that has been bent down and cut by the dropstone. In some places, the downlapping laminations appear to merge with laminae dipping in the opposite direction (vertical arrows). The laminations at the top of the picture are onlapping westward (horizontal arrow) onto the topmost downlapping lamina.

Day 6. Upper Smalfjord Formation on Gæssenjar'ga

Introduction

This day is largely concerned with outcrops at the northern end of Gæssenjar'ga, best reached by boat from Torhop, at the southern end of Tanafjord (Figs. 3 and 46). (If no boat can be obtained from Torhop, the inhabitants of Vestertana should be asked.)

Diamictites, banded diamictites, and deformation diamictites are exposed, the latter two showing glaciotectonic deformation, as well as glaciomarine and interglacial sediments. From correlations with the Vaddasbak'ti area, on Laksefjordvidda (Fig. 3; Føyn and Siedlecki, 1980), these rocks are thought to form the upper part of the Smalfjord Formation, younger than the sediments deposited in the Varangerfjord paleovalley described in previous days.

For a better understanding of the geology of Day 6, Stop 5.3 should be visited first, if it was not already done.

Stop 6.1: View of Gæssenjar'ga from the Road Corner West of Torhop

Location. Drive north from Tana Bru, past Smalfjord and Indre Torhop. Park well clear of the sharp (and dangerous) corner that brings Tarmfjord into view, and climb the small knoll to the west-northwest of the corner.

Figure 45. Geaidnonjoasčohkat, South Tanafjord. All at Stop 5.7. All Thin Submember, Middle Member, Mortensnes Formation. (A) Pod of pale-buff–colored dolomicrite with dolostone clasts, possibly formed by in situ melting of a sunken ice block. (B) Similar pod as in A, but with fewer clasts; note that the sediments on the north side (arrowed) onlap onto the dolostone pod. (C) Loose block showing a glaciotectonic fold in sediments of the basal part of the Middle Member cut by massive diamictite, possibly with a thin deformation diamictite (arrowed). Dashed lines outline the fold. The whole block is cut by an oblique Caledonian cleavage. (D) Typical appearance of the marine sediments of the Thin Submember of the Middle Member. See text for details. (E) Possible dump structure; see text for details. (F) Complex structure, with a possible dropstone having downlap by the sediments on the right (east) side, with later onlap above.

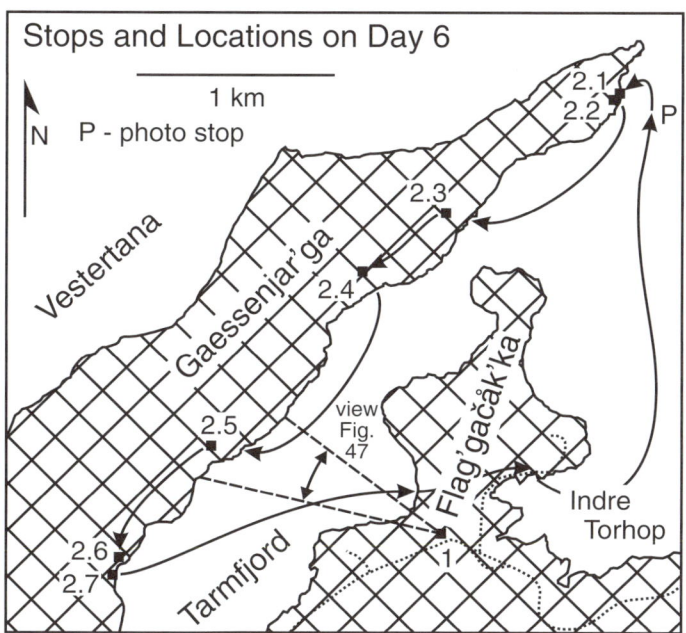

Figure 46. Map showing the location of outcrops visited during Day 6.

Grid ref. NU 36301990, 1:50,000 map sheet Smalfjord 2235 I, Edition 3-NOR. GEC 70°28′56″N, 27°58′18″E.

Introduction. The Smalfjord Formation in the Tanafjord area consists of five cyclical depositional sequences (in the model of Edwards, 1984; Fig. 7), unconformably overlying the Tanafjord Group. From this vantage point, the complete succession, with its variations in color and thicknesses can be seen.

Description. Although the section viewed lies a few kilometers to the north of that shown in Figure 8, the geology is the same as at Addjagiidčåkka. The lower part of the profile (Fig. 47) consists of white quartzitic sandstones of the Dakkovarre Formation (Tanafjord Group; Figs. 4 and 5). The contact with the overlying Smalfjord Formation is not visible but likely lies not far above the uppermost quartzites in the picture. All five members can be seen, although only the lower few are visited during the excursion. The most complete section is on the north side of the gully on the left of the picture. Here, the whole profile, from the upper part of the basal diamictites of Member b through to and including the diamictites of Member e, is visible. From this profile, most of the contacts between the members can be followed across the picture. Member e forms the skyline in the central part of the picture, but it is not visible to the north. Diamictites of Members b, c, and e are buff colored, reflecting a dolomitic component in the matrix. In contrast, those of Member d are greenish brown.

Member a is poorly exposed in Figure 47; generally, it is hidden in the trees on the slope above the Dakkovarre Formation. However, on the south side of the picture, red laminites at the top of Member a can be seen under buff diamictites of Member b (arrowed in Fig. 47).

Glaciotectonic thrusting within Member b is reflected by a number of en-echelon, north-"dipping" gullies, shown as dashed lines in Figure 47.

Stop 6.2: Gæssenjar'ga

Location. From the parking place, drive back toward Indre Torhop and drive down to the harbor to meet with whomever an arrangement has been made for the short boat trip from Tarmfjord to the northernmost point of Gæssenjar'ga (Fig. 46).

Grid ref. NU 3716 2239, 1:50,000 map sheet Langfjord 2236 II, Edition 2-NOR. GPS 70°30′08.0″N, 27°59′46.7″E, near the north end of Gæssenjar'ga.

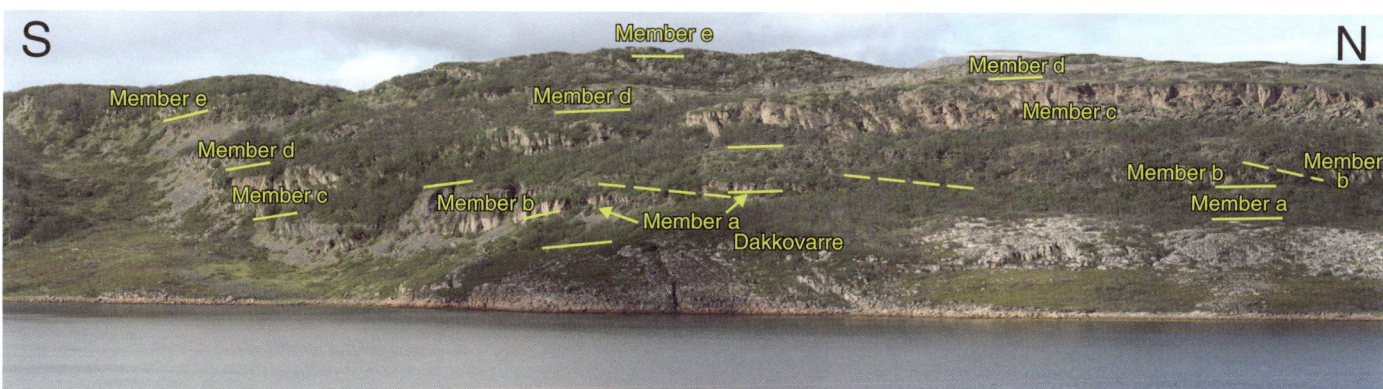

Figure 47. Stop 6.1. West of Torhop. Panorama view across Tarmfjord showing the unconformity between the Dakkovarre and Smalfjord Formations. The field of view is shown in Figure 46. All five members of the Smalfjord Formation can be seen. Of the first glacial cycle (Member a), only the red laminites at the top of the succession are visible, but for the other cycles (Members b–e), both the basal diamictites and the overlying finer-grained successions are visible. However, distinguishing between the laminites at the top of each cycle and the deformation diamictites at the base of the overlying cycle is not possible. The main diamictites of all the members, except for Member d, are buff colored. The valley at the south end of the panorama affords an almost complete profile through the succession. Within the diamictites of Member b, there are a number of north-dipping zones of trees or gullies. These represent planes along which glaciotectonic imbrication of the member occurred (dashed lines in figure). Coast length in view is ~0.5 km.

Introduction. The day involves getting into and out of the boats several times.

Life jackets must be worn properly fastened each time you get into the boats.

The upper part of the Smalfjord Formation consists of a succession of deformation and banded diamictites that Reading and Walker (1966) and Edwards (1984) interpreted to be the result of deformation in the underlying rocks by movement of a glacier, overlain by massive diamictites and thence by finer-grained lithologies with lonestones and dropstones, giving a cyclical sequence (Fig. 48) repeated five times (Members a–e; Fig. 7). Essentially, the succession is composed of alternations of subglacial and subaqueous glacigenic deposits, within which deformation and banded diamictites typically comprise the basal part of the former. In contrast, Hansen (1992), while regarding the succession as being of glacial origin, did not envisage cyclical glacial processes (Fig. 49). An alternative, debris-flow interpretation for the diamictites (Arnaud and Eyles, 2002, 2004) should be kept in mind when looking at the rocks, although it is not alluded to further here.

Description. The description is based on a N-S traverse (Fig. 46). Although several features are documented here, for which GPS coordinates are given, there are very many other outcrops of interest that can be studied and discussed by participants.

Location 1. GPS 70°30′8.0″N, 27°59′46.7″E. The boat should arrive at white to gray quartzitic sandstones with pervasive cross-bedding and thin (0.5 cm) interbedded yellow-greenish silts of the Dakkovarre Formation (Tanafjord Group). Just before landing, the boats can be stopped, to take photographs of the obvious large recumbent fold in the cliffs of diamictite.

Moving into the small bay immediately to the north, the Tanafjord Group is overlain by a dark deformation diamictite (Member a; Fig. 7), including slabs of red-purple sandstone up to 2 m long and 12 cm thick (Fig. 50A) that can be reliably correlated with the Ferruginous sandstone member near the top of the Dakkovarre Formation (seen at Stop 5.3). These are overlain by laminated siltstones and shales and then by red laminated glaciomarine sediments forming the top of Member a. This is overlain by buff-weathering dolomitic diamictite at the base of Member b, comprising a mixture of laminated silts reworked from Member a and dolomitic tillite; essentially a glaciotectonic shear/deformation diamictite (banded diamictite) produced by grounded ice (Fig. 50B).

In the red diamictite from the overlying basal part of Member b, some clasts have a "strain shadow" of different-colored diamictite, sometimes asymmetrically disposed and giving a shear sense (Fig. 50C). The banding here might be due to glacial deformation (M.B. Edwards) or it might be a primary sedimentary feature (T.A. Hansen). There is an absence of primary sedimentary structures; no grading, ripples, or lamination has been seen. The many clasts of dolostone, often pisolitic (Fig. 51A) or stromatolitic, were most probably derived from the Grasdal Formation deposited to the south of Grasdalen (Fig. 3). Just above the seaweed level, there is a large folded clast of reddish shale, with thin paler layers, in the buff-colored diamictite of Member b (Fig. 51B). Translucent bright-red siliceous clasts at this spot are likely also derived from the Grasdal Formation (Fig. 51B, arrowed); similar silicified red shales and dolostones are exposed in situ in the Grasdal Formation.

Location 2. GPS 70°30′7.2″N, 27°59′45.7″E. Not far south of the landing point, uphill and at the contact of the green diamictite and red laminite with gray-buff shales, there is a well-developed fold in alternating silts and diamictite, at the margin of a 12-cm-sized clast (Figs. 51C and 51D), to the south of which is

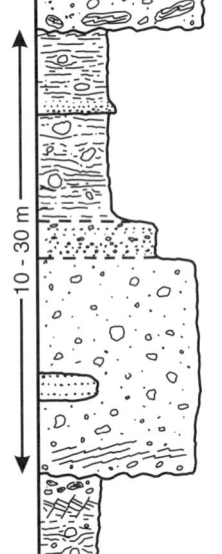

LITHOLOGY	INTERPRETATION
Rhythmically laminated mudstone, with occasional dropstones and rare graded beds.	Deposition from suspension by freshwater overflow plumes in glacial marine setting; also rafting from floating ice, and rare density underflows depositing turbidites.
Crudely, horizontally laminated and bedded, poorly sorted sandstone with pebbles and silt.	Deposition at ice margin by tractional underflows and sediment gravity flows.
Massive diamictite with rare in-situ bodies of stratified sediment.	Lodgment tillite with rare subglacial stream deposits.
Banded diamictite.	Banded lodgment tillite.
Erosion surface.	Subglacial erosion and incorporation of substrate into glacier base.
Deformed substrate.	Subglacial deformation of substrate.

Figure 48. Schematic model for the cyclical deposition of the five members (a–e) within the Smalfjord Formation on Gæssenjar'ga (after Edwards, 1984).

a stromatolitic clast in the green/buff diamictite. Nearby, there are several other glaciotectonic folds, on both a large scale (Fig. 51E; visible from the boat before landing) and a small scale (Fig. 51F).

South of this, at the same height, there is a sedimentary injection vein, up to 1 cm wide and ~70 cm long, with a bridge structure (Fig. 52A). It seems likely that the intrusion formed by fluid overpressures at the ice base and that the sediment-laden fluid was injected downward. Continuing southward, past several more folds, a south-facing joint surface shows a large block of laminite partially incorporated into the banded diamictite with large dolomite clasts (Fig. 52B). Note that the layers here are also cut by a planar injection vein that dies out both upward and downward; this is probably an edge section through a larger elliptical mode 1 fracture surface.

From here, return to the boats and go by sea to the next location; inform the persons in the boat that they should meet the excursion a further 0.5 km to the south of Location 3.

Location 3. GPS 70°29′48.4″N, 27°58′20.8″E. Above a 5 m exposure gap over the Dakkovarre Formation, buff dolomitic diamictite (Member b) contains large blocks of stromatolitic dolomite, derived from the Grasdal Formation (Fig. 52D). Large (5 m × 0.5 m) rafts of red shale/silt, with some lonestones, lie within the diamictites (Fig. 52E). These were derived from the upper part of the underlying glacial sedimentary cycle (Member a).

From here, walk 0.5 km southward, to the next location.

Location 4. GPS 70°29′38.6″N, 27°57′40.1″E. Here, thin sandstones with a sharp lower boundary lie within and grade rapidly up into, red laminites of Member a. These overlie diamictites, below the hammer in Figure 52C. Next, walk south into the valley and look back at the view to the north, showing the overall distribution of Members b and c (Fig. 53); small-scale isoclinal folds are common within the laminites near the south end of the buff diamictites of Member b.

From here, take the boats to the next location. On leaving the boat, inform the persons in the boat that they should meet the excursion at Location 7.

Location 5. GPS 70°29′09.6″N, 27°56′25.8″E. In a small vegetated notch up within the Smalfjord Formation, spectacular glaciotectonic normal faults dipping roughly northwestward cut a raft of red and greenish-brown laminite within a gray, slightly buff diamictite. Both of these were derived from Member c and form the deformation diamictite at the base of Member d (Fig. 52F). The angle of dip of the faults is too low for them to be typical Andersonian normal faults, and their orientation is

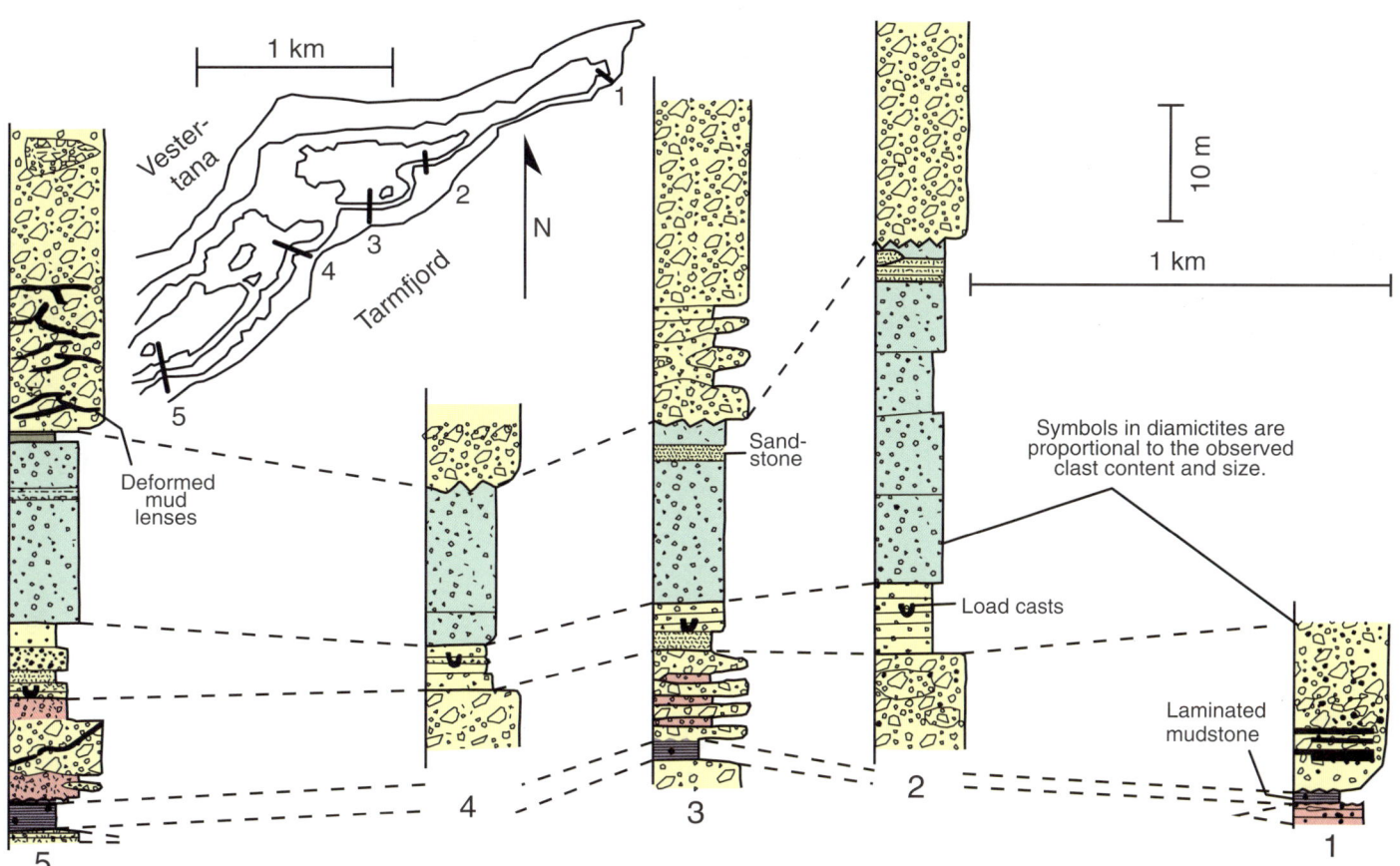

Figure 49. Logged-sections through the Smalfjord Formation in the northern part of Gæssenjar'ga (after Hansen, 1992). Profile 5 is from essentially the same place as the Addjagiidčåkka profile of Figure 8.

Figure 50. Gæssenjar'ga. All at Stop 6.2. All Smalfjord Formation. (A) Location 1. Panorama view of Members a and b, overlying the Dakkovarre Formation quartzites. (B) Location 1. Folded shale-diamictite contact, Member b. (C) Location 1. Low-strain diamictite adjacent to a clast in Member b. Note that the sedimentary boundary directly above the clast and the boundary below it (passing at the top of the coin) are hardly deflected around the clast, indicating that little compaction (or flattening) occurred. Note also the different color of the diamictite in the strain-shadow area. The strain shadows (arrows) are slightly asymmetric with respect to the glaciotectonic banding, suggesting shear with a top-to-the-north (right) sense. This is only one example and is not consistent with the regional picture.

not consistent with the regional Caledonian or post-Caledonian faulting directions. Possibly the fractures represent asymmetric boudinage. These are overlain by diamictites with centimeter-scale shale clasts (derived from the Vagge Formation, Tanafjord Group) of Member d.

The "fault rock" (Fig. 52F) overlies a dolomitic diamictite (Member d) with stromatolite clasts and clasts of buff-brown diamictite from Member c (Fig. 52G). Note that there are large variations in the texture and color of this stratigraphic unit, due to the fact that it was created from mixing two very different sediment populations. The material in Figure 52G could have been formed from 90% red muds and 10% dolomite, in which case it would not here have the typical buff color of dolomitic diamictite.

Stratified glaciomarine diamictite, with some possible grading (Fig. 54A) and ripples in the lower part, can be seen in the upper part of Member c.

Moving northeast for up to 100 m, several superb dropstones in red glaciomarine rocks at the top of Member c can be seen in the cliff exposures (Figs. 54B and 54C).

From here, walk south to the next outcrop.

Location 6. GPS 70°28′51.1″N, 27°55′41.1″E. Superb glaciotectonic deformation (Fig. 54D) can be seen in the central part

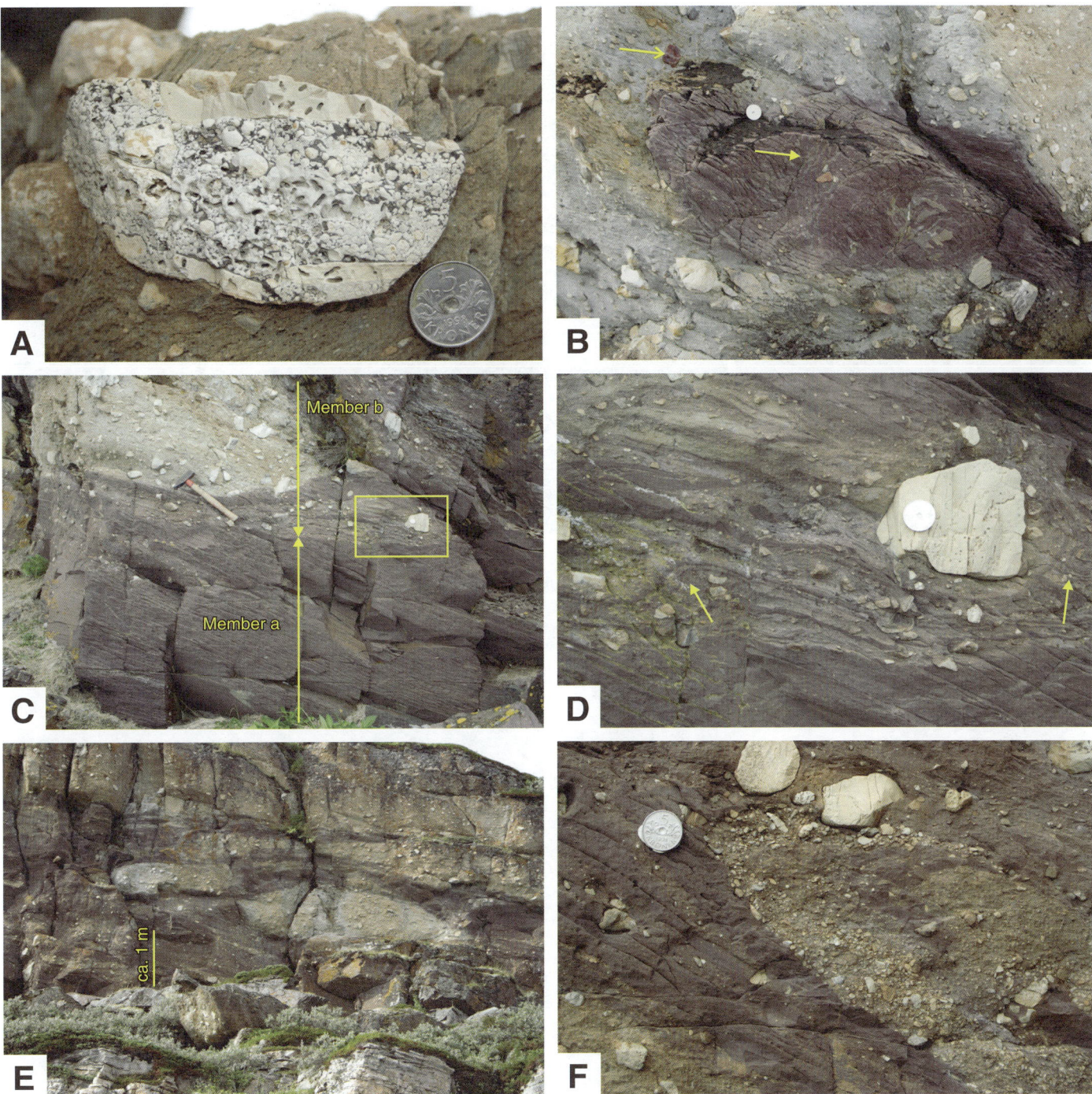

Figure 51. Gæssenjar'ga. All at Stop 6.2. All Smalfjord Formation. (A) Location 1. Close-up of dolostone clast from the Grasdalen Formation, formed from pisolites and edgewise breccia fragments of micrite, in diamictites of Member a. (B) Location 1. Folded (solid arrow) red laminite clast from Member a in dolomitic diamictites of Member b. Open arrow points to a siliceous red clast derived from the Grasdal Formation. (C) Location 2. Contact of Member a laminite and overlying Member b diamictite. Area of part D is shown. (D) Location 2. Enlargement of part C, showing folds developed in banded diamictite of Member b adjacent to a dolostone clast (arrowed). (E) Location 2. Large-scale fold in banded diamictite of Member b; this can be clearly seen from the boats when approaching the first location. The red material was derived from the laminite of Member a. (F) Location 2. Small-scale fold in diamictites of Member b.

Figure 52. Gæssenjar'ga. All at Stop 6.2. All Smalfjord Formation. (A) Location 2. Injection vein of relatively pale-buff–colored sediment cutting across the boundary of Member a laminite and the overlying deformation diamictite at the base of Member b. The vein almost certainly continues upward, out of the plane of view. Open arrow points to a fold; solid arrow points to a left-stepping bridge structure in the injection vein. (B) Location 2. Partial incorporation of red laminite of Member a into the overlying Member b deformation diamictite. Note clast of stromatolitic dolostone (arrowed). (C) Location 4. Thin white weathering sandstones within laminites of Member a. (D) Location 3. Inverted large clast of stromatolitic dolomite derived from the Grasdalen Formation, in diamictite of Member b. Coin (in box) for scale. (E) Location 3. Raft of red laminite from Member a in diamictites of Member b. (F) Location 5. Glaciotectonic faults cutting a raft of Member c laminite in Member d diamictite. (G) Location 5. Buff diamictite clast of Member c within basal red diamictites of Member d.

of a 1-m-thick clast of interbedded red and gray laminite within red diamictites of Member c. The clast was derived from the glaciomarine facies at the top of Member b.

The upper and lower laminations in the clast are undeformed. In contrast, in the central part, disharmonic folds, thrusts, and flanking structures have all developed. The strain intensity dies away gradually downward and, more importantly, upward. There is no evidence for erosion of the deformed layers prior to deposition of the overlying undeformed layers, indicating that deformation occurred after all the sediments were deposited. This is comparable to deformation at the base of submarine slumps, which also dies away from the detachment surface. In this interpretation, the deformation that caused the folding was caused by movement of an overlying glacier; subsequently, the whole clast was extracted from its depositional place and incorporated into the diamictite. Above this, the contact between the diamictites and the laminites can be seen, with the former "digesting" the latter, forming comparable fold structures near the contact.

From here, walk south to the next locality.

Location 7. GPS 70°28′48.5″N, 27°55′44.2″E. This last outcrop also shows the glaciomarine facies of Member c, with abundant dropstones within a sequence of sometimes graded beds (Fig. 54E). Some clasts of buff diamictites (probably from Member b) may be present. Unusual beds up to 6 cm thick with abundant clasts (up to 3 cm sized) are present within a sequence of finer-grained beds (Fig. 54F) with lone/dropstones. Such beds indicate either mass-flow deposition or periods of intense rainout of clasts from floating ice. The former seems more likely but is not certain.

From here, return to the boats, cross directly over the fjord and walk over the hill to Torhop.

Day 7. Mortensnes Formation in the Las'sasuolo-Låk'sunjåkka Coastal Profile, West Tanafjord

Introduction

The Mortensnes Formation, the upper glacigenic unit in Finnmark, and its relationship to the unconformably underlying Nyborg Formation, forms the dominant focus of the day. In this area, the Lower Member is absent and the Thick Submember forms the base of the Mortensnes Formation (Fig. 9). Structures within the Nyborg Formation and the overlying Lillevatn Member of the Stappogiedde Formation are also seen.

The Mortensnes Formation has been correlated with the 582–584 Ma Gaskiers glaciation, based on the extremely negative $\delta^{13}C$ isotope values (–7.57‰ to –9.91‰; Wonoka anomaly, Halverson et al., 2005) in dolostones of Member E of the Nyborg Formation underlying the Mortensnes Formation in the Trollfjord-Grasdal area (Figs. 3, 4, 6, and 7).

Stop 7.1: Las'sasuolo-Låk'sunjåkka (Stappugied'di) Coastal Profile

Location. Drive to wherever arrangements have been made to take boats across to the island of Las'sasuolo and thence northwards along the coast (Fig. 55). The Larsen family, which lives in the Vestertana area, can be contacted through the Tana Bru Kommune offices about this.

Life jackets must be worn properly fastened each time a boat trip is made.

Figure 53. Gæssenjar'ga. Stop 6.2. Smalfjord Formation. Location 4. Panorama view of Member c overlying Member b.

Figure 54. Gæssenjar'ga. All at Stop 6.2. All Smalfjord Formation. (A) Location 5. Graded bedding in diamictite of Member d. (B–C) Location 5. Dropstones of basement and dolostone, respectively, at the top of Member b. Note the graded bedding at the top of C. (D) Location 6. Glaciotectonically folded lamination in a clast of glaciomarine laminites from Member b, within Member c. (E) Location 7. Graded bedding in diamictite with dropstones, Member c. (F) Location 7. Very coarse-grained conglomeratic bed within finer-grained sandstone beds, probably reflecting a mass flow within Member c.

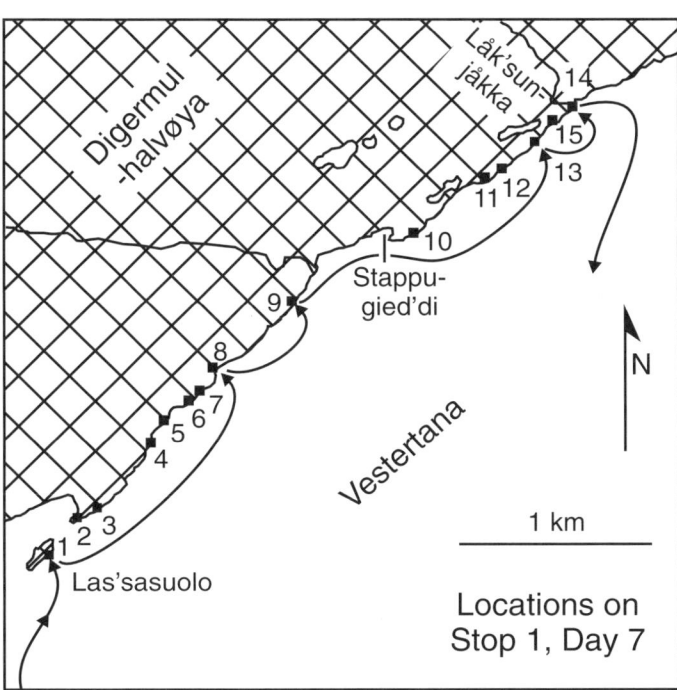

Figure 55. Map showing the location of outcrops visited during Day 7.

If you hire a boat and go without a local person, be aware that Vestertana is dangerously shallow in several places in the general area between Torhop and Las'sasuolo. Ask a local person about the best route.

Grid ref. NU 70°31'36.31"N, 27°58.05.62"E (east side of Las'sasuolo).

Introduction. Sufficient details of the Mortensnes Formation have been given in the Geological Introduction and at Stop 5.7. Figure 56 summarizes most of the important details for the profile examined here, which corresponds to column 5 in Figure 9. Although the trip passes from the base to the top of the formation when traveling from south to north, the strike of the bedding is essentially subparallel to the coastline in this area. However, the succession is cut by several NW-SE–striking normal faults, downthrowing to the north.

Description. The description given is based on a south to north traverse in the boats. Many outcrops are documented in the following section, for which GPS or GEC coordinates are given, but there are a great number of other outcrops of interest. The water is deep by the rocks, so one can go in close to examine them and to take photographs.

Location 1. GEC 70°31'36"N, 27°58"E. Cross Vestertana to the southeast side of Las'sasuolo, where it can be seen that glacial erosion at the base of the Mortensnes Formation cut a channel

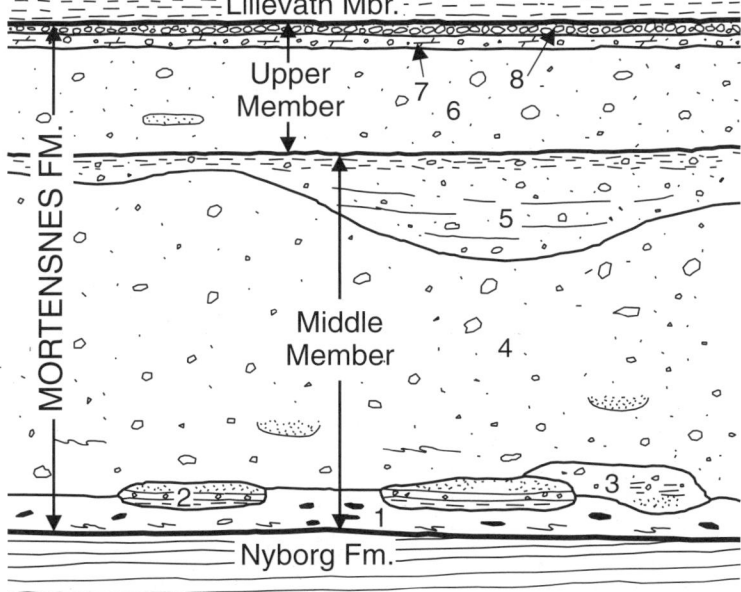

Unit 2 - large tabular blocks, many of which include Nyborg rafts, diamictite and white sandstone.
Unit 1 - blanket of massive deformation tillite; purple-gray-green. Mostly Nyborg detritus; rare dolomite and crystalline clasts.

Upper Member
Unit 8 - thin polymict lag conglomerate.
Unit 7 - thin dolomitic diamictite with small dolomitic clasts and occasional large basement clasts.
Unit 6 - sharp-based, locally erosive, deformation diamictite at base with overlying gray diamictite, with crystalline clasts (5%-15%), most showing rounding, in muddy matrix with scattered quartz grains.

Middle Member
Unit 5 - bedded diamictite similar in color and composition to Unit 4. Erosive base. Some conglomerates. Overlain by laminite (sand/mud with numerous dropstones). In-situ sandstone bodies show soft-sedimentary deformation.
Unit 4 - buff-brown tillite with very sandy matrix, mostly dolomite clasts, with many sedimentary and deformation structures (folds, thrust & normal faults). Many oval clasts of paler dolomitic sandy diamictite, lying parallel to bedding. Also clasts of dolomitic breccia. Light weathering sandstones, as wisps and lenses, fill channels.
Unit 3 - stratified dolomitic diamictite (10%-30% clasts, of dolomite & chert) and sandstone; observed at only a few localities. Diamictite has a dolomitic matrix with scattered sand grains.

Figure 56. Schematic illustration of the stratigraphic units within the Thick Submember of the Middle Member and the Upper Member of the Mortensnes Formation in the Tanafjord area (after Edwards, 1984).

in the underlying Member C of the Nyborg Formation and filled this with buff-weathering diamictite of the Middle Member (Fig. 57A). Only a very thin deformation diamictite is present. Clasts of diamictite in diamictite, called "ghost" clasts, formed from a slightly different buff-colored diamictite and up to 10 cm across, are exposed here (Fig. 57B). The source of the ghost clasts was an underlying diamict that was either frozen together or had undergone rapid carbonate cementation prior to reworking.

Location 2. GPS 70°31′42.7″N, 27°58′24.9″E. On the mainland, near the southern tip of the small peninsula, more ghost clasts are seen in the diamictite overlying Member C of the Nyborg Formation. Moving north, an ~1-m-size clast of edgewise breccia, derived from the Grasdal Formation, occurs in the Middle Member of the Mortensnes Formation, characterized by a variably developed but always present buff-weathering color (Fig. 57C). Banding, showing glaciotectonic folding, occurs in the Middle Member here. Further on, the succession enters purple-colored rocks of Member D of the Nyborg Formation (Fig. 7).

Location 3. GPS 70°31′44.3″N, 27°58′34.7″E. This outcrop was originally interpreted by Edwards (1972) as a channel-structure in green turbiditic rocks of the Nyborg Formation (Member C), subsequently filled with purple-colored tidal rocks of Member D (Fig. 57D). An alternative, and now preferred, interpretation is that the Member D rocks were pulled out of position by the base of a glacier and dragged along, imbricating the upper part of the deeper-water facies rocks of Member C into a duplex; they are, thus, part of a large, glaciotectonically complex clast of the Nyborg Formation, now forming the base of the Mortensnes Formation.

A 20-cm-thick pale sandstone showing hummocky cross-stratification confirms the deeper-water nature of the underlying sediments of Member C of the Nyborg Formation (Fig. 58A). In this area, only a thin (0.5 m) layer of deformation tillite is present

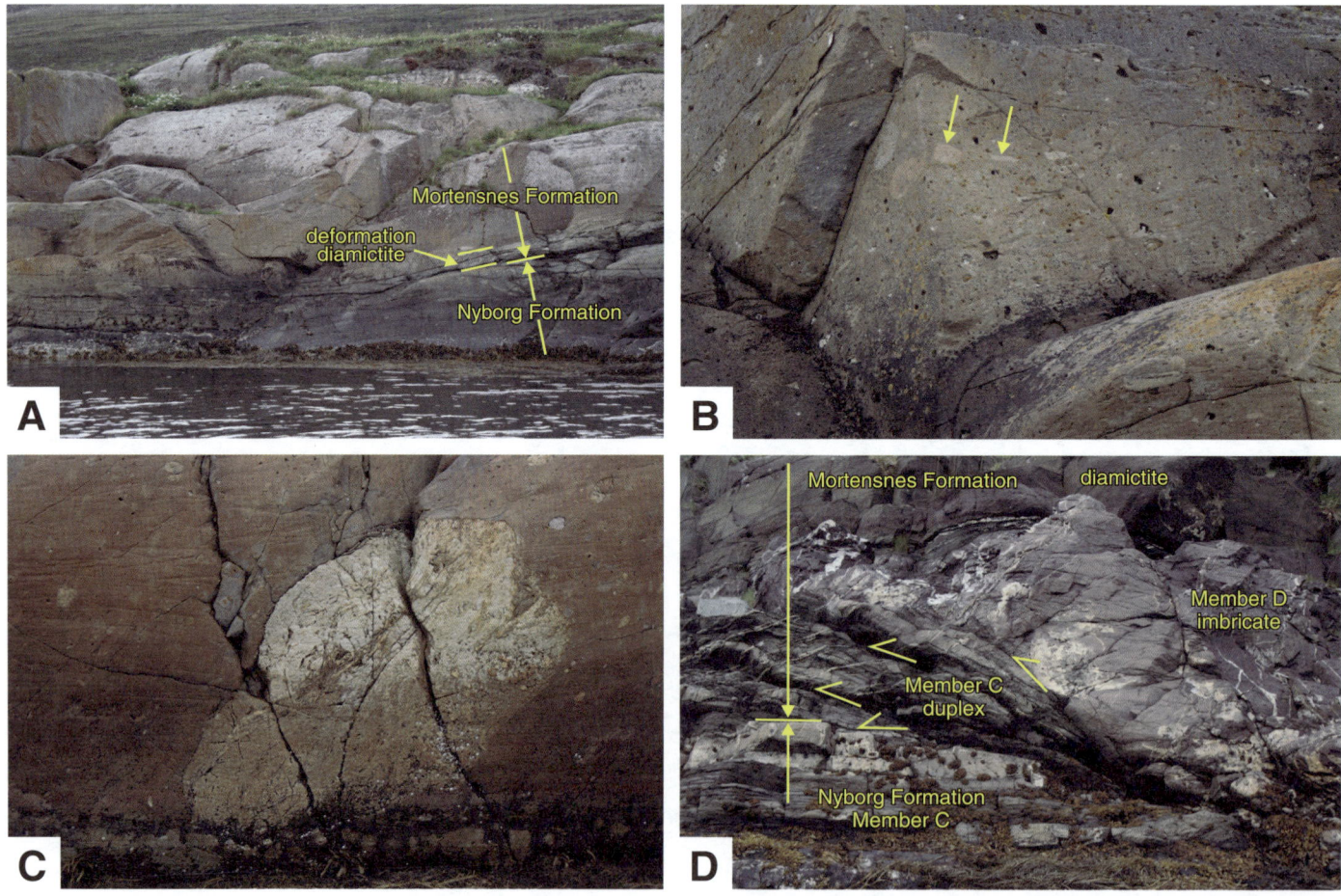

Figure 57. Stappugied'di profile. All Stop 7.1. (A) Location 1. Diamictite of the Mortensnes Formation (Middle Member, Thick Submember), filling a channel in Nyborg Formation, Member C. The basal part comprises a deformation diamictite. (B) Location 1. Ghost clasts (arrowed; ~10 cm in size) in the Mortensnes Formation. (C) Location 2. Very large (~1 m) and isolated edgewise breccia clast in diamictites of the Mortensnes Formation. (D) Location 3. Glaciotectonically imbricated rocks of the Nyborg Formation (thus forming the base of the Mortensnes Formation). A duplex of small imbricates derived from Member C is overlain by a large displaced imbricate of purple-colored rocks of Member D. These are overlain by diamictites.

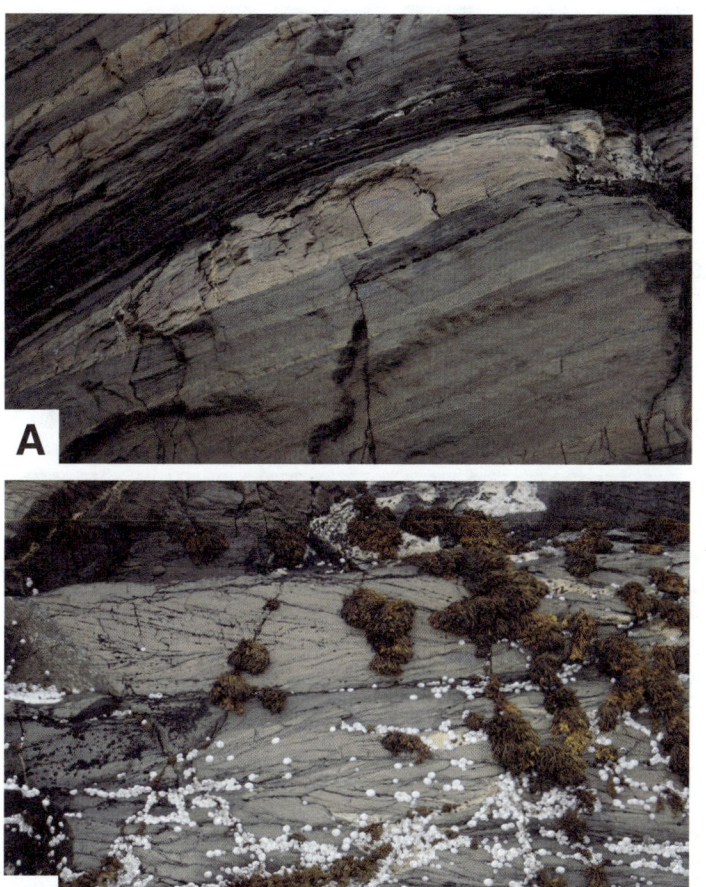

Figure 58. Stappugied'di profile. All Stop 7.1. (A) Location 3. Hummocky cross-stratification in the Nyborg Formation (Member C). Sandstone ~25 cm thick. (B) Location 3. Thin (~0.5 m) deformation diamictite derived from Member C of the Nyborg Formation. (C) Location 6. Herringbone cross-bedding in Member C of the Nyborg Formation.

at the base of the Mortensnes Formation (Fig. 58B). This is gradationally overlain by the typical orange-buff diamictites of the Middle Member.

Location 4. GPS 70°31′55.6″N, 27°59′03.3″E. Slightly further north, a 10-cm-thick sandstone channel in the Mortensnes Formation is exposed.

Location 5. GPS 70°31′59.0″N, 27°59′08.4″E. North of the previous location, ghost clasts are common in the orange-weathered Middle Member diamictites. Note that the bedding in Member C of the Nyborg Formation here is thinner than to the south, indicating this locality lay further down the paleoslope.

Location 6. GPS 70°32′02.4″N, 27°59′19.5″E. Herringbone cross-bedding is exposed here (Fig. 58C), overlain a few meters north by convolute bedding structures in Member C of the Nyborg Formation. These are overlain by deformation diamictites forming the base of the Mortensnes Formation.

Location 7. GPS 70°32′04.3″N, 27°59′26.8″E. Thin (30 cm) red sandstones in green, more pelitic rocks of Member C of the Nyborg Formation show superb convolute bedding structures (Fig. 59A); this lenses out rapidly along strike.

Location 8. GPS 70°32′08.3″N, 27°59′34.2″E. In good weather, the participants can land here on the northeastward-facing beach. The bay shows a profile from Member C of the Nyborg Formation, on the east side of the headland, before coming to the beach, through all the Mortensnes Formation, to the Lillevatn Formation on the west side of the bay. The Middle Member is relatively thick here; it thins to the south. At the top, it is finely laminated with clasts of pisolite and laminated dolomite (Fig. 59B). The source of such clasts was probably the Grasdal Formation, although to the south, this is thought to have been eroded away by the Smalfjord Formation. The lower part of the Upper Member is gray-green colored, with basement clasts. The top is a buff-orangey weathered diamictite, with a dolomitic matrix. The Lower Submember of the overlying Lillevatn Member is thin here, with a rusty to sulfur-yellow weathering (Fig. 59C).

Some 50 m north of the bay, a complex contact between the Nyborg and Mortensnes Formations crops out (Fig. 59D). The upper part of the Nyborg Formation, here Member C, has been informally divided into four parts, from the base upward.

Figure 59. Stappugied'di profile. All Stop 7.1. (A) Location 7. Convolute bedding structures in a ~30-cm-thick red sandstone in the Nyborg Formation, Member C/D. (B) Location 8. Pisolite clast from the Grasdal Formation, in the top part of the Middle Member, Mortensnes Formation. Note the strain shadow around and pressure-solution cleavage within the clast. Solution of the overlying quartz-rich clast has also occurred at the clast-clast contact. (C) Location 8. Lower Submember of the Lillevatn Member of the Stappogiedde Formation. (D) Location 8. Contact of the Nyborg and Mortensnes Formations. The Nyborg Formation here consists of four units: Unit W is the normal Member C deposits; Unit X is a sandstone layer showing convolute bedding due to dewatering; Unit Y is the sand that was expelled during dewatering, containing large fragments (arrowed); and Unit Z is a thin layer of cross-bedded light-colored sandstone. The overlying base of the Middle Member consists of the stratigraphy outlined in Figure 56. A basal deformation diamictite (Unit 1) is overlain by a complex succession of Unit 2, documented in Figure 61. The raft of Nyborg Formation (Unit 2a) is from Member C. Ed. '72 shows the area of Plate 134 in Edwards (1972). Schematic hammer for scale.

Unit W consists of normal lithologies of Member C; Unit X has convolute bedding structures, formed during dewatering of the underlying sediments; Unit Y is a massive, structureless sandstone, the product of dewatering, with fragments of ball-and-pillow structure derived from Unit X; and Unit Z is a narrow zone of white, cross-bedded sandstones deposited in channels in Unit Y and likely formed by currents sorting and reworking Unit Y. Figure 60A gives a close-up view of the upper part of the Nyborg Formation, showing that Unit Z is persistent at the top of the Nyborg Formation (a detailed photograph of the Unit Z is given in Edwards, 1972, his Plate 134).

These are overlain on a sharp, sheared contact (Fig. 59D) by Unit 1 of the Mortensnes Formation (cf. Fig. 56), this being a deformation diamictite mostly derived from the Nyborg Formation but with rare dolomite and crystalline basement clasts. Above is Unit 2, with a raft of allochthonous Nyborg at the base (Unit 2a), including beds that have internal soft-sediment deformation, and proglacial sand bodies at the top (Unit 2c). Unit 2b (diamictite) is not clearly visible, but the rounded outcrop in Figure 59D may be this part of the succession.

Location 9. GPS 70°32′19.8″N, 28°00′20.2″E. Going north, there are some small sandstone-filled channels in the Nyborg Formation (Fig. 60B). Then comes a very large sandstone body. This was interpreted by Edwards (1972) as having been formed proglacially, deposited in a channel carved into earlier deposits of the Middle Member of the Mortensnes Formation (Figs. 60C and 61). A number of features can be seen here, so land in the bay to the north, if the weather is suitable.

Figure 60. Stappugied'di profile. All Stop 7.1. (A) Location 8. This shows in detail the contact between the Nyborg and Smalfjord Formations. See caption to Figure 59D for details. Schematic hammer for scale. (B) Location 9. Thin (~20 cm) sandstone channels in Nyborg Formation. (C) Location 9. Large proglacial channel, filled with ~3.5 m of white sandstones, in the Middle Member of the Mortensnes Formation; terminology of units is from Figures 56 and 61. Arrow points to position of Figure 62D.

The white sandstones filling the channel (Unit 2C; Fig. 61) were most probably derived from the abundant well-sorted sandstones and quartzites in the Tanafjord Group (Fig. 4). The channel is ~3.5 m deep, with a very steep southern channel margin (Fig. 62A), reflecting the strength of the frozen tillite as it was eroded. The sandstones were steeply banked up against this surface, onlapping it, as the channel was infilled. Minor interfingering of the sandstones and diamictites occurs near the base of the steep wall of the channel (arrowed), and, just above this, the sandstones appear to have collapsed, forming very small-scale, almost isoclinal folds as well as a breccia, in the sediments (Fig. 62B, arrowed). In the diamictite underlying the northern edge of the channel, a relatively well-developed Caledonian cleavage formed adjacent to the sandstones, probably due to the marked change in rheology across the contact.

The northern end of the channel lies at a slightly lower angle (Fig. 62C) and also shows interfingering of the sandstones and underlying diamictite (arrowed); the bedding at the base of the channel here climbs up the channel margin and continues as a much thinner layer for ~5.5 m to the northeast.

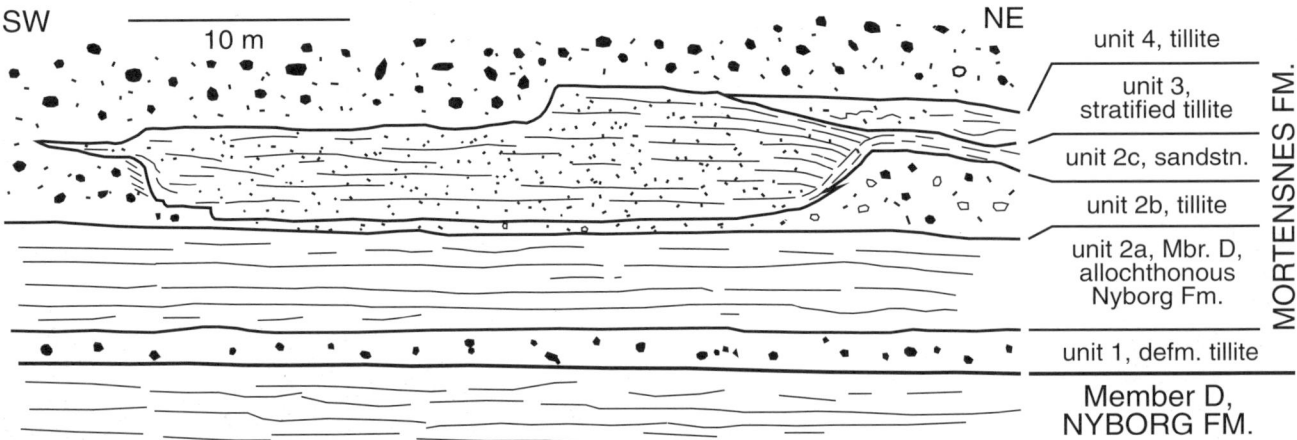

Figure 61. Sketch of lithologies at the large sandstone body seen in Figure 60C (from Edwards, 1972). See Figure 56 for a description of the units indicated.

Higher sandstone beds within the channel onlap these lower beds toward the north.

Near the top of the main channel, at the northern end, a prominent thick white sandstone bed within the channel is sharply offset, while the stratigraphy of the thin-bedded sandstones immediately below varies in thickness and detailed lithology across the fault. This suggests that postdepositional thrust faulting (glaciotectonic) occurred below the thin beds (Fig. 62C), dropping the thick bed down, together with another, overlying thin-bedded succession that is not seen in the footwall. The stratigraphy above and below this "package" is unaffected by the steep fault (a lateral ramp, essentially).

Above Unit 2c, at the northern end, there is a stratified diamictite deposited in open subglacial channels (Unit 3; Fig. 61). This shows convolute lamination and reworked, likely frozen, clasts of diamictite (Fig. 62D). These deposits may have formed from diamictite that slumped into the open river channel below the glacier. Alternatively, they may have formed as pulse deposits that failed, either due to traction forces or during dewatering.

Low-amplitude and essentially symmetrical, somewhat sinuous ripples (or ripple-like structures) were found by Camille Partin, in a pale gray-blue sandstone within the stratified tillite sediments a few tens of centimeters above a large white sandstone lens within Unit 3 (Fig. 62E). These sandstones were likely formed by reworking of the directly adjacent, similarly colored diamictites. Interpreting these "ripple" structures is very difficult and is not hazarded here. Small-scale conglomerate-filled channels also occur within Unit 3, reflecting cyclical contemporary erosion and deposition within the subglacial deposits (Fig. 62F).

Below the main sandstone channel-fill (Unit 2c), there is a thin (~30 cm) layer of dolomitic diamictite with clasts derived from the Grasdal Formation (Unit 2b) and then a massive raft of allochthonous Member D of the Nyborg Formation within the Mortensnes Formation (Unit 2a). This is underlain by a deformation diamictite (Unit 1). Since there is no sign of glacial deformation to indicate displacement between this massive raft and the sandstone lens, the latter must also be allochthonous.

After examining the main sandstone lens, participants should scramble around or over the minor headland to the beach outcrops to the south. Here, Nyborg Formation–derived deformation diamictite at the base of the Mortensnes Formation (Unit 1) records subglacial deformation of the Nyborg Formation substrate (Fig. 63A). This overlies Member D of the Nyborg Formation, containing ripples and convolute bedding structures (Fig. 63B).

In the white sandstones (Unit 2c) above this deformation diamictite, glaciotectonic imbrication of the white sandstone indicates a southerly ice flow (Figs. 63C and 63D). Both small-scale glaciotectonic normal and thrust faults developed in carbonate-rich layers within the white sandstone (Fig. 63E), as well as folds (Fig. 63F); altogether, there is a lot of evidence for glaciotectonic medium- to small-scale deformation. On a larger scale, there are two further white sandstone layers above the main sandstone outcrop described (Fig. 63D). These may reflect larger-scale glaciotectonic imbrication of a single sandstone body.

There is a spatial relationship between the presence of sandstone bodies and the large rafts of Nyborg at the base of the Middle Member (Edwards, 1984), strongly suggesting that the sandstone lenses were deposited on only a thin layer of diamictite above the top of the Nyborg Formation. Although melting must have occurred, allowing the proglacial channel to form, subsequent freezing, during/after deposition of the overlying diamictite, welded the relatively massive sandstones to the underlying Unit 1 diamictites and Nyborg Formation, so that they were imbricated together along a weak (likely a shale) horizon within the Nyborg Formation. The deformation diamictite lies at the base of the sequence, and so was the last diamictite to form. Edwards (1984) proposed that the subglacial channels (Unit 3) formed after glaciotectonic imbrication.

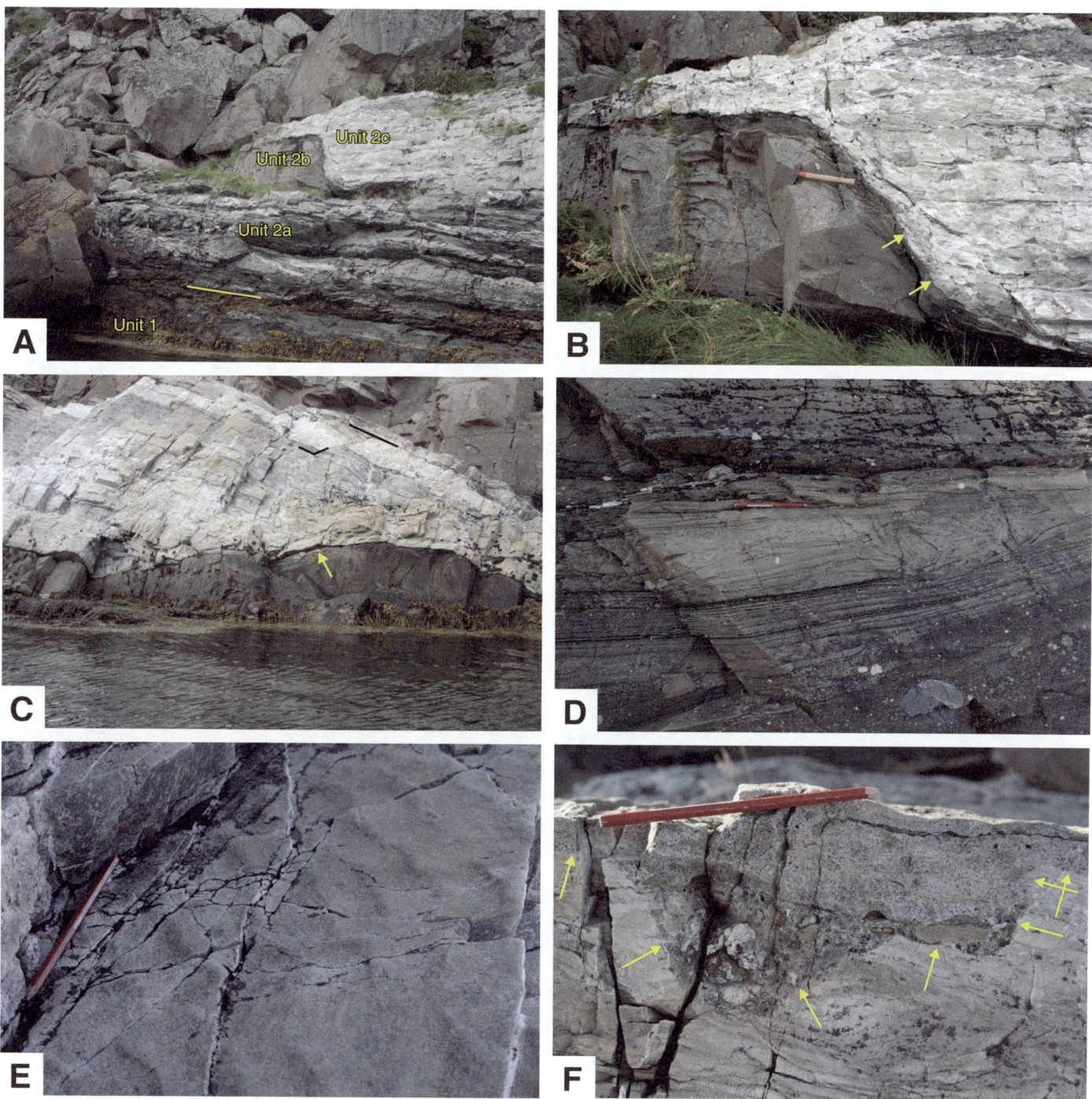

Figure 62. Stappugied'di profile. All Stop 7.1. All Location 9. All Middle Member, Mortensnes Formation. (A) Overview of the southwest end of sandstone channel, showing Units 1, 2a, 2b, and 2c. (B) Close-up of the southwest end of the large Unit 2c sandstone channel, showing interlayering of sandstones and diamictite and small-scale isoclinal folding and brecciation of sandstones (both arrowed). (C) Close-up of the northeast end of the large sandstone channel. Note the interlayering of diamictite and sandstone at the base of the channel (arrowed) and the offset of the thick white sandstone layer near the top of the channel fill (outlined). White sandstones altogether are ~3.5 m thick. (D) Instability structures in the Unit 3 stratified diamictite, deposited in a subglacial channel. See Figure 60C for locality. (E) Possible symmetrical ripples in the Unit 3 stratified diamictite, at the northeast end of a large sandstone channel. (F) Channel within the sandstones of Unit 3; arrows delineate channel geometry.

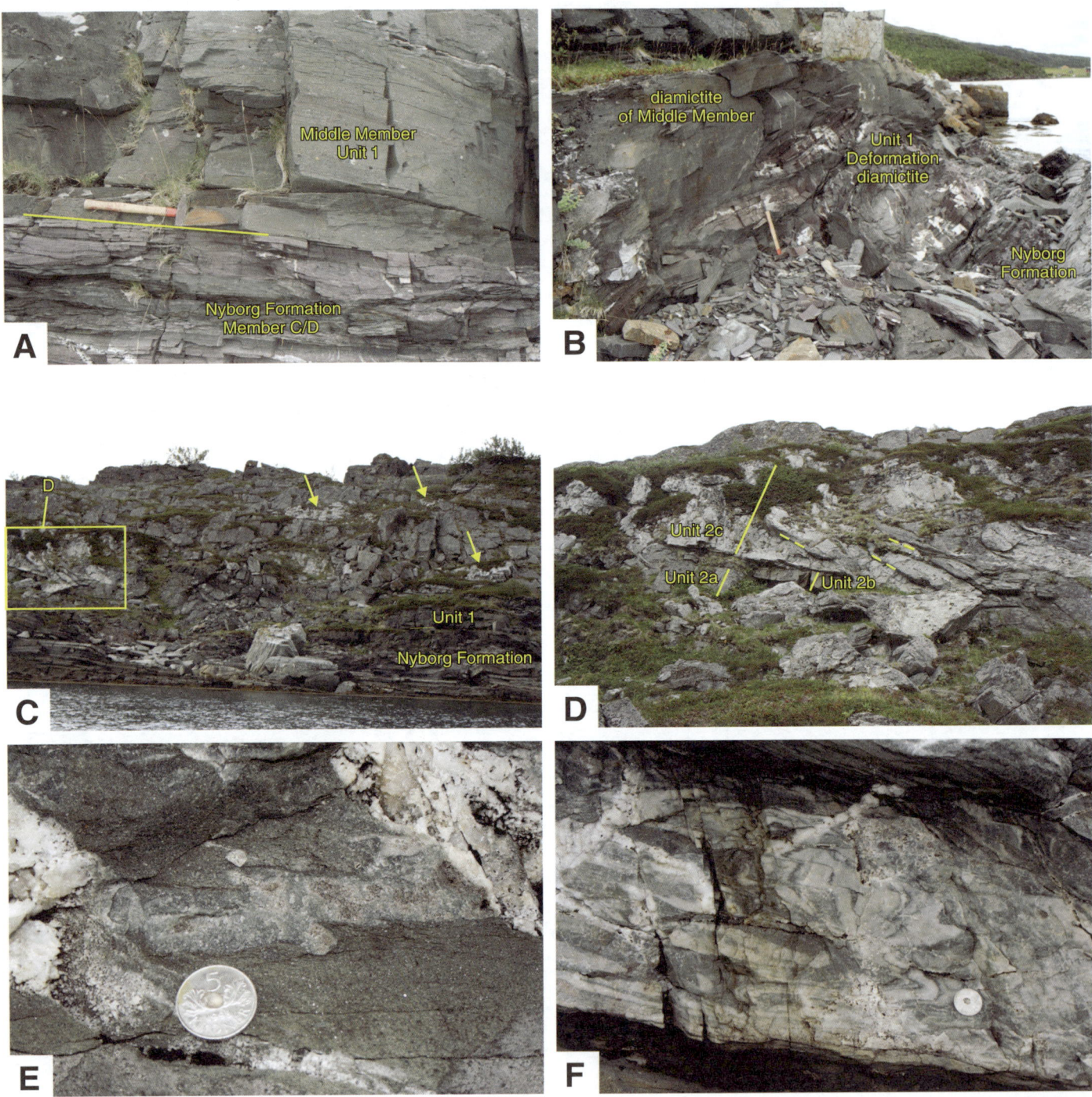

Figure 63. Stappugied'di profile. All Stop 7.1. All Location 9. All Middle Member, Mortensnes Formation. (A) Nyborg Formation overlain by deformation diamictite of the Middle Member (Unit 1). (B) Transition of in situ Nyborg Formation through Unit 1 deformation diamictite to massive diamictite. (C) View from offshore showing three levels (maybe glaciotectonic imbricates) of proglacial sandstones (Unit 2c; arrowed) and the position of D. (D) Succession from Unit 2a to Unit 2c; the latter has been shortened by glaciotectonic imbrication (dashed lines show possible thrusts). (E) Small-scale thrusting in Unit 2c sandstones in the area of D. (F) Small-scale folding in Unit 2c sandstones in the area of D.

Just north of this outcrop, a major northward-dipping post-Caledonian normal fault is crossed, with Stappugied'di farm in the bay slightly further to the north. The Lillevatn Member is exposed on the west side of the bay.

Location 10. GPS 70°32′30.4″N, 28°01′16.1″E. Here, channels up to 2.5 m thick in diamictite crop out. These are filled with deformed, alternating moderately sorted sandstones and diamictite (Fig. 64A); the diamictite may have periodically slumped in from the banks or been washed down into the channel. Somewhat further north of the Caledonian/post-Caledonian normal fault, well-developed glaciotectonic folds and thrusts can be seen in the Middle Member, which here is thicker and has abundant platy, streaked-out clasts, indicating that they were soft during glaciation.

Location 11. GPS 70°32′40.0″N, 28°01′50.6″E. At this point, several huge asymmetric sandstone lenses, deposited in proglacial channels, crop out (Fig. 64B). Some 30 m to the north, a large boulder of brecciated Grasdal Formation lies within the Middle Member, with some carbonate material passing into the surrounding diamictite as a result of reworking of the sediment (Fig. 64C).

Slightly to the south of this locality, grooves can be seen in the top of Unit 4, under Unit 5, in the Mortensnes Formation (cf. Plate 173 *in* Edwards, 1972).

Location 12. GPS 70°32′42.4″N, 28°01′58.8″E. More sandstone bodies are exposed here, within the Middle Member, together with southeast-facing soft-sediment folds in the sandstones (Fig. 64D).

Location 13. GPS 70°32′47.0″N, 28°02′16.2″E. This is a tiny cleft in the coast, just wide enough for one boat; if the weather is suitable, the excursion should land. Immediately north of the landing point, a huge clast of dolostone that appears to have undergone a complex deformation history lies within the Middle Member (Fig. 65A). On a large scale, it has been folded, as outlined in Figure 65A using a photograph taken in the early 1970s, when less debris overlay the structure (Fig. 65B).

In the top left of Figure 65A, a wedge of gray rock surrounded by darker buff-colored diamictite may be a pinch-and-swell

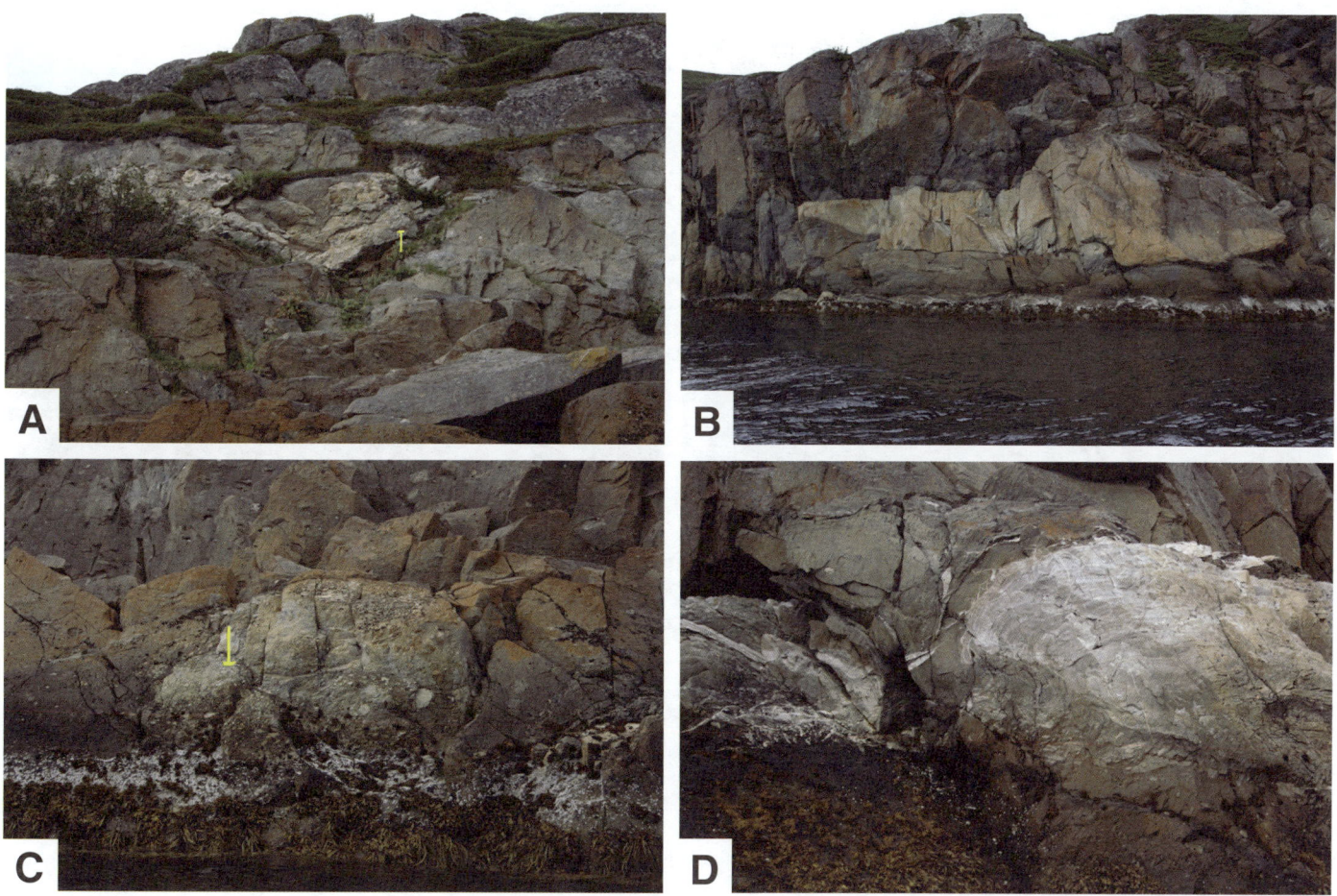

Figure 64. Stappugied'di profile. All at Stop 7.1. All Middle Member, Mortensnes Formation. (A) Location 10. Folded proglacial sandstones interbedded with diamictites. Schematic hammer for scale. (B) Location 11. Large asymmetric proglacial sandstone body. (C) Location 11. Clast of dolomitic breccia derived from the Grasdal Formation. Schematic hammer for scale. (D) Location 12. Deformed sandstone lens.

structure within the overall large-scale clast (Figs. 65A and 65C). The contact between the gray and dark buff layers, and less so between the dark and light buff layers, has been offset on low-angled top-to-west extensional planes at several places, possibly in association with the formation of what seems to be a ?pressure solution cleavage that is refracted at the contact of the darker buff and surrounding lighter buff carbonate (Fig. 65C).

Near the sea, a 30-cm-sized asymmetric "clast" (in structural terms a σ-clast) wrapped by the sedimentary layers indicates dextral movement (top-to-southeastern quadrant; Figs. 65A and 65D). The thinning of the layers underneath this clast, in a high-strain area compared to the low-strain area to the right, confirms that the dolostone was not cemented at the time of deformation.

We acknowledge that this deformation history is speculative, but it should form the basis for a more detailed investigation by future excursions.

Paul Hofmann noted that the deformation must have been soft-sediment deformation, since the temperature for ductile deformation in dolomite is ~400 °C, far in excess of the temperatures attained during the very low-grade metamorphism that occurred in the Gaissa Thrust Belt (Rice et al., 1989b).

The source of the dolomite forming the clast is unknown; the three possibilities are the Grasdal Formation and Members A and E of the Nyborg Formation. Both of the latter two are very thin and are unlikely to have contributed much detritus to the glacial sediment. Thus, the most likely source of the dolostone is the Grasdal Formation, although this was strongly eroded during the Smalfjord Formation glaciation (Marinoan). No δ^{13}C data are currently available from this outcrop.

Location 14. GPS 70°32′53.3″N, 28°02′34.4″E. The excursion should land here, on the south side of a small river, at this, the northern end of the profile (not on the rocky beach 30 m north). Just north of the river, the thin buff-weathering dolomitic

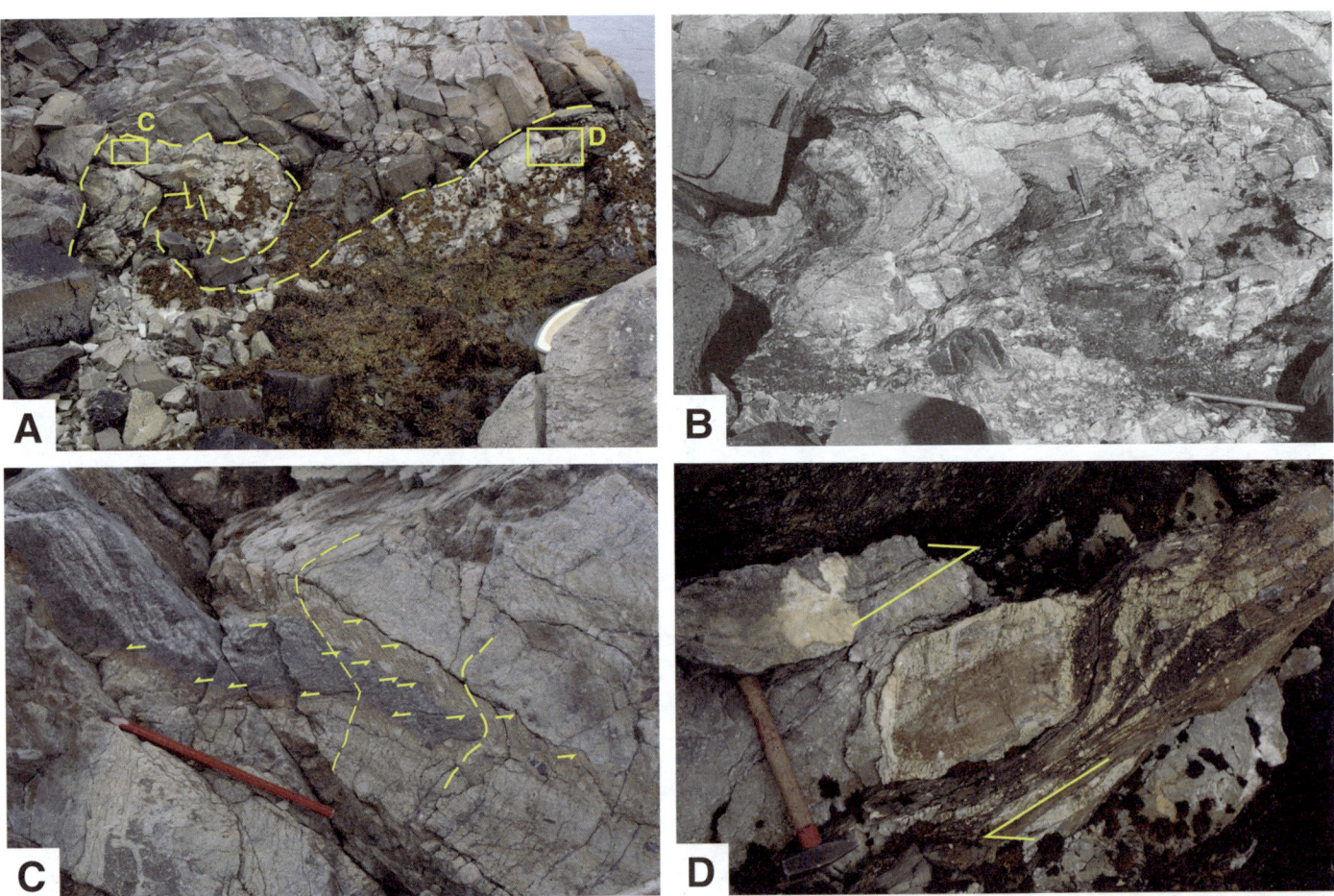

Figure 65. Stappugied'di profile. All at Stop 7.1. Location 13. All Middle Member, Mortensnes Formation. (A) Large deformed clast of dolostones; this is of uncertain origin, but the folding is not a result of Caledonian tectonics, and so the carbonate must have been uncemented when folded. Coarse dashed line outlines the fold shape. Positions of C and D are arrowed. Schematic hammer for scale. (B) Close-up of part A. This black-and-white photograph was taken in the early 1970s by Marc Edwards, when the structure could be seen better. (C) Close-up of the northwest part of the fold showing the refracted ?cleavage (dashed lines) and the sheared offset margins of the lithologies (shear half-arrows). (D) Dolostone σ-clast indicating top-to-the-right (east) shear. Note thinning of the glaciotectonic compositional banding as it goes around the clast.

diamictite forming the top of the Upper Member of the Mortensnes Formation is exposed (Fig. 66A). This contains small, mostly rounded dolostone clasts, less than 3 cm in size, including oolite clasts, and occasional subrounded clasts of basement up to 30 cm in size, in a slightly sandy and silty but generally fine dolomitic matrix (Fig. 66B). This unit has a relatively sharp, planar lower contact with the underlying tillites of the Upper Member, which has abundant crystalline basement clasts and is much less dolomitic than either the Middle Member or the top of the Upper Member. The lower part of the Upper Member is thought to be a ground moraine, while the upper part is likely a marine deposit.

Four matrix samples from the upper part of the Upper Member from here gave $\delta^{13}C$ values between –8.42‰ and –10.44‰ (VPDB; Rice and Halverson, 2005–2010, personal observations), strongly suggesting derivation from the dolostone found close to the top of the Nyborg Formation (Member E; Figs. 4, 6, and 7) in the Trollfjord area; this has been correlated with the Wonoka anomaly (Halverson et al., 2005). No oolites have been recorded in Member E, but they are present within the much older Grasdal Formation.

To the east, a shallow gully leads down to the sea; the lower left of Figure 66C shows essentially undeformed diamictite of the basal part of the Upper Member—a crystalline clast-rich diamictite, without a buff matrix. This merges upward into a carbonate-rich and carbonate-poor banded diamictite, with abundant small carbonate clasts. The banding is glaciotectoic in origin; this also caused both isoclinal and low-angle reverse faulting (both arrowed in Fig. 66C) and later open folding.

The lack of augen structures around the clasts seems strange, considering they have been seen at other localities, but, again, this is not unusual for banded diamictites generally.

Closer to the sea, the rocks show more intense deformation (Fig. 66D) in buff-colored carbonate-rich layers and blocks of carbonate-poor material (black). In some layers (arrowed), the color is as intense as in pure dolostone. Note the folds and refolds

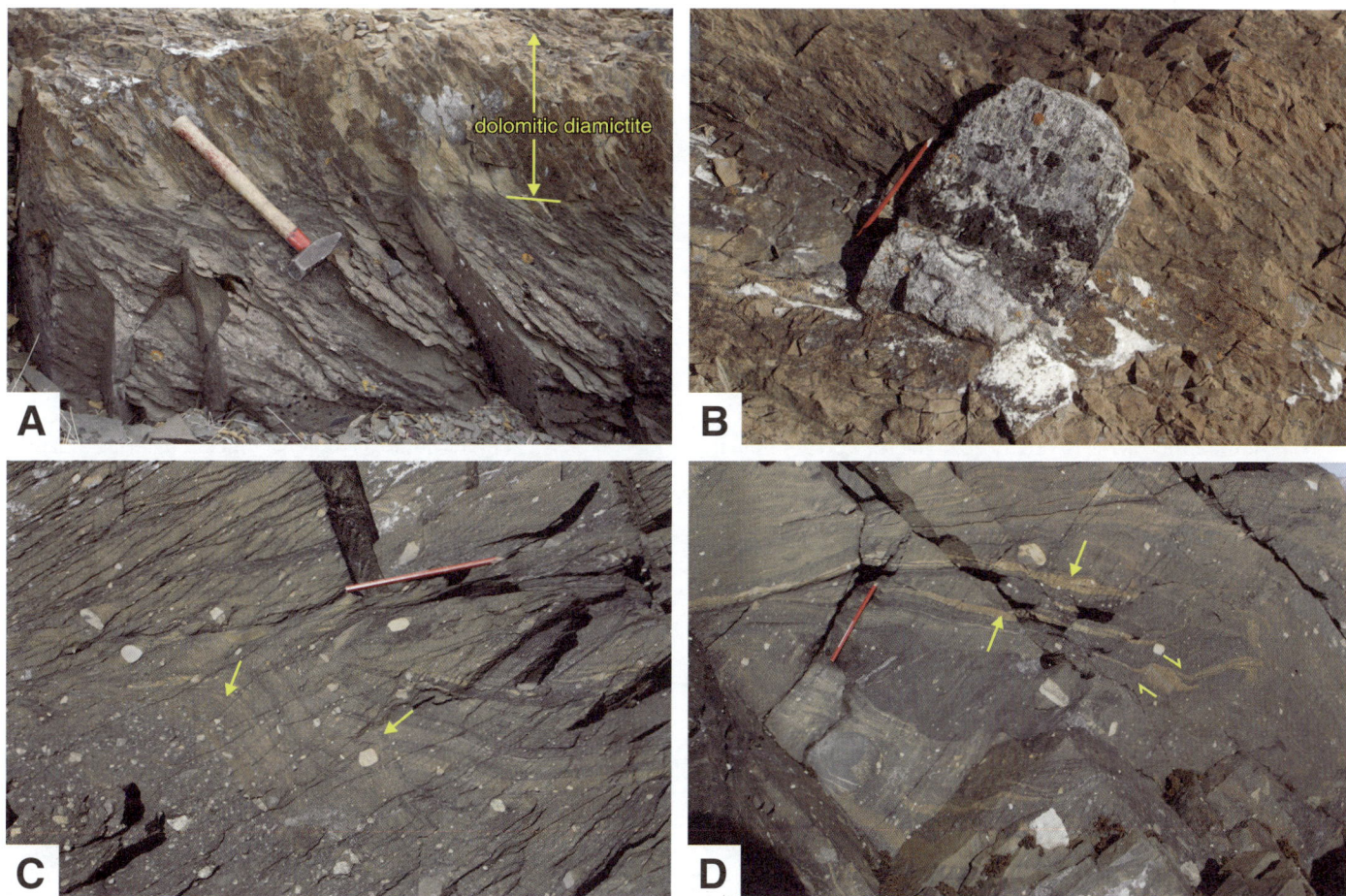

Figure 66. Stappugied'di profile. All at Stop 7.1. Location 14. All Mortensnes Formation. (A) Buff-weathered dolomitic tillite at the top of the Upper Member. (B) Large basement-derived and smaller dolostone clasts within the dolomitic tillite of the Upper Member. (C) Transition from massive diamictite of the Upper Member to deformed banded diamictite, at the gradational boundary from the Middle to Upper Members. Arrows point to glaciotectonic features; on left to an isoclinal fold within sediments and on right to a thrust fault. (D) Deformation diamictite at the base of the Upper Member, derived from the Middle Member. Note σ-clast indicating top-to-the-right (east) shear (half-arrows) and bands of essentially pure dolostone (arrowed).

and the sigma-clast (Fig. 66D), very similar to those seen at the previous location; the latter shows the same sense of shear (Fig. 65D).

In summary, glacial retreat occurred at the end of the deposition of the glaciomarine rocks at the top of the Middle Member. A subsequent advance, with ground-based ice, led to the deposition of lodgement till (lower part of Upper Member). This gave way to the glaciomarine deposits seen in the upper part of the Upper Member.

Location 15. GPS 70°32′50.1″N, 28°02′25.0″E. From the sharp bend in the river at the south end of the pebble beach at Location 14, walk SW for ~70 m, along the west side of the low hillock forming the outcrop adjacent to the coast, and then turn southeast toward the sea.

On the outcrops facing the sea, a long (>12 m) and thin (maximum 60 cm) lens of laminated subglacial sandstones and gravels ca. 2 m above the base of the Upper Member crops out. Along a sharp boundary (Figs. 67A and 67B), the base of the sandstone consists of a unit with mud flakes, indicating high energy, and also some lenses of dark diamictite, presumably transported while frozen. Several dykes of these sediments have been injected downwards from this layer, into the underlying diamictite, suggesting high fluid overpressures, and then ptygmatically folded, forming lobate-cuspate structures (Fig. 67B), probably during compaction. The cusps point into the sandstones, indicating that this was the more competent lithology. The lithology in the veins gradually becomes finer grained downward, diverging from a composition identical to that in the source rock.

Day 8. Associated Neoproterozoic Strata along the Varanger Coast (Vadsø to Hamningberg)

Introduction

The aim of this day is to briefly examine rocks older than the Smalfjord Formation that may be relevant to correlations of the Neoproterozoic rocks in Finnmark with other successions worldwide. The locations of the outcrops are shown in Figure 68.

Drive east, along the north coast of Varangerfjord. Initially, the exposures are alternations of the Nyborg Formation and the glacial units, due to large-scale folding. East of the museum at Mortensnes, the road goes past essentially horizontally bedded, siliciclastic rocks of the Vadsø Group (Figs. 3 and 4), frequently with abundant sedimentary structures. The first outcrop documented here lies slightly east of the airport on the east side of Vadsø.

Stop 8.1: Sjåbuselva, East of Kiby

Location. Take the dirt road ~80 m east of Kiby and drive north for ~1.3 km (Fig. 68A and inset). Park off the road in the obvious large clearing on the right (east) and walk northward near the river, to north of the remains of concrete weirs. The outcrops are in the steep west bank of the river, above the rusty-weathering white sandstones in the river.

Grid ref. PT 0932 7685, 1:50,000 map sheet Vadsø 2435 III, Edition 2-NOR. GEC 70°04′19″N, 29°52′23″E.

Introduction. There are no known glacial rocks below the Smalfjord Formation; evidence of an earlier Cryogenian ("Sturtian") glacial event is, therefore, absent, unlike many other Neoproterozoic successions (cf. Arnaud et al., 2011). Whether this is because the Sturtian glaciation, for which there is a wide range of isotopic ages (741–643 Ma; cf. Fairchild and Kennedy, 2007), was not a worldwide event or because it was subsequently eroded away in Scandinavia, is unknown. In Laurentia, it commenced at 717 Ma, while in the Nubian Shield, it persisted to 711.5 Ma (Bowring et al., 2007; Macdonald et al., 2010). This reflects either a diachronous glaciation or the overall length of the event. In the sub-Smalfjord sequence, the most significant unconformities, which might be "hiding" a glacial event, lie at the base and top of the Lille Molvika Formation (Ekkerøya Group); the former, exposed at this locality, is more significant in terms of a time gap (Figs. 4 and 5; Siedlecka, 1995).

Figure 67. Stappugied'di profile. All at Stop 7.1. Location 15. All Mortensnes Formation. (A–B) Flattened injection veins of sandstone into diamictite of the Upper Member, about 2 m above the boundary of Units 4 and 5 (Fig. 56). In B, note the lobate-cuspate geometry of margin of the veins (arrowed) and the change in color of the margin of this vein where it cuts through the topmost dark shaley part of Unit 4.

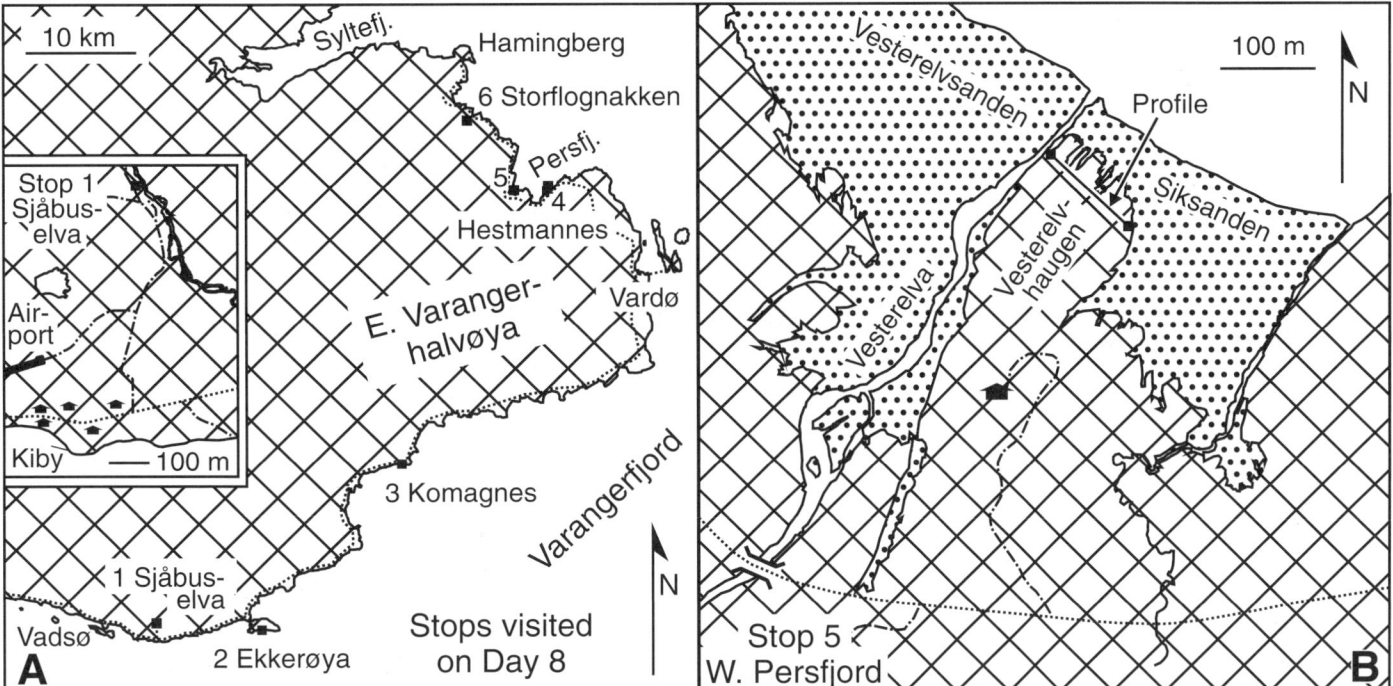

Figure 68. (A) Map showing the location of outcrops visited during Day 8. Dashed lines indicate roads. Inset shows location of Stop 8.1. (B) Detailed map showing the location of the profile visited for Stop 8.5.

At the outcrop here, the Lille Molvika Formation overlies the Vadsø Group on a very slight angular unconformity. In the Manjunnas area of west Varangerhalvøya (Fig. 3), the Lille Molvika Formation overlies rocks of the Barents Sea Group (combined Båtsfjord and Tyvjofjell Formations; Fig. 5), comparable to, but not exactly the same as, those in the area north of the Trollfjorden-Komagelva Fault, along an angular unconformity, with a sedimentary breccia at the base (Figs. 3 and 5; Rice, 1994). Thus, there is a marked break in the sub–Lille Molvika Formation sedimentary facies between the Manjunnas area and this stop.

Description. Over ~30 m along the river, the Golneselva Formation (top Vadsø Group; Fig. 4) has an irregular upper surface, with ~1.8 m of paleorelief. The base of the overlying Lille Molvika Formation is a 0.25–2-m-thick sedimentary mud-flake breccia with subangular to subrounded sedimentary clasts in a conglomeratic matrix. Breccia clasts are soft, platy, up to 0.4 m in size, and randomly distributed and oriented. Angular 1-cm-sized K-feldspar clasts are also present; thus, both a proximal basement and a reworked cover source are implied. The matrix content varies, with up to 50% in places, although the rock is probably mostly clast supported. The matrix is a greenish rust-spotted sandstone with subrounded, subspherical to subelongate quartz clasts up to 3 cm size (Fig. 69A). This breccia is irregularly overlain by a 25–30-cm-thick bedded layer without or with very small breccia clasts, and this is then overlain by 90 cm of breccia similar to the base of the section, although the clasts are subparallel to bedding. Bedding in the conglomerate is slightly discordant to the bedding in the underlying Golneselva Formation. Upstream, a 1.5 m × 0.75 m "block" of Golneselva Formation sandstone lies in the lower breccia. This may be reflecting an undercut paleotopography rather than a sedimentary clast, although Banks et al. (1971) reported even larger clasts at the base of the Lille Molvika Formation elsewhere.

Stop 8.2: Ekkerøya Bird Colony

Location. Return to the main road and drive east for ~8.5 km. Turn south off the main road, toward Ekkerøy; although "-øya" indicates an island in Norwegian, Ekkerøya is connected to the mainland by a narrow sandy isthmus, much favored on sunny days by the local people, since one or the other side is usually wind-free. Continue into the village and turn left, uphill, at the sign to the bird colony. Park at the end of the road and walk east along the base of the cliffs.

The old fish factory in the village is sometimes open as a café, with traditional fishing equipment displayed.

Grid ref. UC 9050 7648, 1:50,000 map sheet Ekkerøy 2435 II, Edition 3-NOR. GEC 70°04′13″N, 30°07′37″E.

Introduction. Although the Lille Movika Formation is underlain and overlain by unconformities, it has a similar facies to the Vadsø and Tanafjord Groups, indicative, overall, of rapidly changing sea levels, to give a thick shallow-marine–deltaic–fluvial succession deposited in this boundary area between the Gaissa and the Timan basins (cf. Siedlecka et al., 2004).

Description. The Lille Movika Formation consists of interbedded green-gray mudstones and siltstones, pale-gray sandstones, and subordinate conglomerates (Siedlecka and Roberts, 1992). Mudstones are more abundant toward the base of the

Figure 69. West Varangerhalvøya. (A) Stop 8.1. Conglomerate of large siltstone/mudstone clasts within sandstones at the base of the Lille Molvika Formation, Kiby, near Vadsø. (B) Stop 8.5. Contact between the Båsnaering (right, south) and Båtsfjord (left, north) Formations in Persfjord. The person is pointing to where a thin basal conglomerate was found. Yellow lines show bedding strike directions; note the ~5° angular difference between the two units.

succession; the sandstone-dominated parts, higher up, contain ball-and-pillow structures and slump folds. Røe (1975) and Johnson (1975, 1978) interpreted the sequence as showing an initial transgression followed by regression within a deltaic or coastal interdeltaic environment.

The N-S–oriented, east-dipping dolerite dike cutting the sequence has an age of 349 ± 10 Ma (recalculated K-Ar whole-rock age; Beckinsale et al., 1976).

Stop 8.3: West of Komagnes

Location. Drive some 25 km further along the coast, through rocks of the Tanafjord and overlying Vestertana Groups, to where an obvious steeply dipping dolerite dike cuts an old sea cliff on the north side of the road.

Grid ref. VC 0395 9150, 1:50,000 map sheet Ekkerøy 2435 II, Edition 3-NOR. GEC 70°12′25″N, 30°27′29″E.

Introduction. The outcrops here consist of the Innerelv Member of the Stappogiedde Formation; these overlie the thin Lillevatn Member, which, in the Tanafjord area, rests on the Mortensnes Formation (Fig. 4). The road goes over the contact between the Tanafjord and Vestertana Groups somewhere between Ekkerøya and this stop but neither glacial formation has been preserved in this area, although a thin slice of the Mortensnes Formation lies some 10 km to the east, and the Nyborg Formation is also absent (Siedlecki, 1980). This suggests that considerable pre- and some post-Mortensnes Formation erosion occurred, assuming a complete succession was initially deposited in this region.

Description. The Innerelv Member of the Stappogiedde Formation typically consists of blue-green, locally redder toward the base, shales, with parallel lamination and small-scale ripple cross-lamination. Both facies indicate a very quiet-water depositional environment, from either suspension or weak bottom currents (Siedlecka and Roberts, 1992). Banks (1973) suggested a shelf-environment influenced by weak tidal currents. Elsewhere, Ediacaran and other trace fossils have been found in this unit (Banks, 1970; Farmer et al., 1992; Crimes and McIlroy, 1999).

The N-S–oriented, east-dipping dolerite dike cutting the sequence has an age of 360 ± 10 Ma (recalculated K-Ar whole-rock age; Beckinsale et al., 1976). In detail, this is a composite dike, which on a larger scale steps to the right (Rice et al., 2004). At very low tide, the beach section shows an number of minor segments of the dike with complex bridge geometries. A close-spaced, steeply dipping fracture cleavage has developed adjacent to the dike, gradually dying away over several meters from the narrow baked margin. Within the baked margin, fractures are less common, perhaps due to annealing during intrusion or the increased strength of the rock imparted by the baking.

Around the sharp corner from this outcrop, the road crosses a wide bay to Komagvær. Somewhere under the eolian sands lies the Trollfjorden-Komagelva Fault, a dextral strike-slip structure with a very poorly constrained displacement of 200–400 km (cf. Siedlecki, 1980; Rice et al., 1989a).

Stop 8.4: SSW of Hestmannes

Location. Drive north, toward Vardø, but just before going into the tunnel to that town, take the turning to the north, toward Hamningberg. On approaching the area, look for the point where the outcrops between the road and sea suddenly change from being rugged, with subvertical beds of sandstone, to a lower, and at medium to high tides, submerged zone (Fig. 68A). This outcrop should be visited at partial to complete low tide.

Grid ref. VD 1771 1501, 1:50,000 map sheet Vardø 2535 IV, Edition 3-NOR. GEC 70°25′52″N, 30°49′03″E.

Introduction. The rocks north of the Trollfjorden-Komagelva Fault are part of a much thicker sequence; the Barents Sea Group has a maximum thickness of 10 km, and the unconformably overlying Løkvikfjellet Group has a thickness of 5.8 km (Fig. 5; Siedlecka and Roberts, 1992). This difference is "accommodated" by the Trollfjorden-Komagelva Fault (Figs. 2 and 5). To the north, the rocks have been strongly folded with at least two generations of deformation; regional studies showed a dome and basin fold geometry in the east (Roberts, 1972; Siedlecki, 1980), while detailed microstructural studies identified three cleavages in the west (Rice and Frank, 2003). The area deserves serious structural analysis.

Rice et al. (1989a), following the then-prevailing concepts, ascribed all the deformation to Caledonian orogenesis, with a component of passive-roof duplex back-thrusting in the east to give the dome and basin interference folding. In contrast, Roberts and Siedlecka (2002), following the newer prevailing dogma, ascribed some or most of the deformation to the Timanian event (latest Precambrian). In the absence of any isotopic age constraints, either model, or a combination, is at least possible.

Description. Well-exposed, thick-bedded, light-gray to pink quartz-rich sandstones and thin sandstones with thin draping greenish siltstones make up the eastern side of the outcrop. These form the uppermost part of the Hestmann Member of the 2.5–3.5-km-thick Båsnæring Formation. The Hestmann Member (800 m thick here) has been interpreted as a braided delta-plain deposit, terminating the deltaic sedimentation of the Båsnæring Formation (Siedlecka and Edwards, 1980; Siedlecka et al., 1989).

These are overlain by carbonate rocks of the Annijokka Member of the Båtsfjord Formation. The contact is an irregular erosion surface (Fig. 69B) along which rare very thin pockets of a basal breccia, with intraformational shale clasts, have been found. Bedding across the contact shows an ~5° difference in strike. Whether this indicates some contemporary local block-faulting rotation or simply erosion by several meters down into large-scale delta foresets needs to be clarified.

The Annijokka Member consists of interbedded, very dark- to pale-gray clays, shales, thin-bedded gray sandstones, greenish-gray mudstones, and gray, buff-weathering dolostones. There are also subordinate limestones, some with domal stromatolites up to 0.5 m high. The rocks contain abundant flaser and lenticular bedding structures, shrinkage and desiccation cracks, and intraformational breccias (Siedlecka and Roberts, 1992). The sequence has been interpreted as a tidal-flat deposit (Siedlecka, 1978, 1982).

Twelve samples from this section all gave negative $\delta^{13}C$ values (−4.9‰ to −2.0‰, VPDB; Rice and Halverson, 2005–2010, personal observations).

Stop 8.5: Persfjord

Location. Continue to drive toward Hamningberg. Park just before the small wooden bridge over the Vesterelv (GEC 70°25′33″N, 30°43′10″E) and walk NNE along the coast until good outcrops are reached. Then, turn east and traverse the outcrops across the peninsula to the beach or continue around on the coastal outcrops (Fig. 68B). If time is available, the beach can be crossed, and the exposures on the NNE-trending coast, which gradually cuts across the sequence, can be visited.

Grid ref. VD 1537 1496, 1:50,000 map sheet Vardø 2535 IV, Edition 3-NOR. GEC 70°25′44″N, 30°43′30″E.

Introduction. The Skovika Member (1.1–1.3 km thick) continues the carbonate-rich succession of the Båtsford Formation seen at the last stop, although the amount of carbonate decreases upward. Altogether, this is the thickest carbonate unit within the northern Scandinavian "Caledonides." Thirty-seven samples from the succession here all gave negative $\delta^{13}C$ values, and similar values were obtained from samples collected near the top of the formation in Syltefjord (Fig. 2; Rice and Halverson, 2005–2010, personal observations). Although there are no robust age constraints on the age of the Båtsfjord Formation, the occurrence of molar-tooth structures suggests an age of pre–ca. 750 Ma (cf. Shields, 2002).

Thus, a correlation of this thick sequence of negative $\delta^{13}C$ values with the Bitter Springs anomaly, dated to ca. 810 Ma (Fanning et al., 1986; Halverson and Shields-Zhou, 2011), seems likely.

Description. The rocks in this steeply dipping section show the wide range of lithologies present within the Skovika Member. Interbedded violet and green mudstones, gray and pink sandstones, buff-weathering gray dolostones, and pale-gray to black limestones occur. Crinkled, possibly algal, lamination, stromatolites, birds-eye structure, ptygmatically flattened molar-tooth structures, and sheet cracking, with isopachous cement fills, have developed (Siedlecka and Roberts, 1992; Rice, 2000, personal observations). No detailed sedimentological work has been done in this unit, although a shallow-marine to possibly occasionally tidal environment is probable.

The Annijokka Member lies to the west, across the Vesterelv, with the Hestmann Member forming the obvious subvertical outcrops behind.

Stop 8.6: Storflognakken

Location. Drive north to the road cut at the Grid Ref given below; exposures are common, and all show essentially the same features. After inspecting the rocks, it is suggested that excursions drive to the end of the road, at Hamningberg. This used to be a major fishing center, but now no one lives here during the winter. Although several pre–World War II houses remain, the fish factories, which previously crowded the shoreline in the village up to the late 1970s, have mostly collapsed and been removed.

Grid ref. VD 1345 2205, 1:50,000 map sheet Vardø 2535 IV, Edition 3-NOR. GEC 70°29′12″N, 30°40′37″E.

Introduction. The Kongsfjord Formation is interpreted as a flysch sequence that accumulated on a submarine fan (Siedlecka, 1972; Pickering, 1981), giving a succession of at least 3.5 km thickness; the base of this unit, the oldest Neoproterozoic rocks in East Finnmark, is not exposed (Fig. 5). The beds seen here are thought to represent outer-fan deposits that accumulated rapidly.

Description. Steeply dipping to vertical, very thick-bedded, dark-gray turbidites can be seen in road cuts and beach exposures on the road toward Sandfjord. Graded bedding in sets from a few centimeters to over 1 m in thickness (in amalgamated beds) is present along with parallel horizontal-lamination at the tops of the graded units (Siedlecka and Roberts, 1992).

ACKNOWLEDGMENTS

We express our unbounded gratitude to Arild and Jorunn Pettersen, and their sons, of Skiippagurra Vest (Tana, Finnmark) for their hospitality during a dry run of the excursion in 2007, during the International Geoscience Programme (IGCP) Project 512 and the 33rd International Geological Congress (IGC) excursion itself in 2008, and in *very* many of the previous summers since 1990.

The staff at Europcar Norway (Kirkenes) are thanked for their patience as the size of the original excursion changed daily and for sponsoring the car transport used during the excursion. Else-Marie Seim, headmistress of the school at Seida, is very sincerely thanked for allowing the excursion participants to use the school facilities in 2008. The Tana Kommune office at Tana Bru, especially Teresa Nyborg, is thanked for their help organizing the boat rides, and Mia Krogh at Varangerbotn Museum is thanked for providing information about the Pomor trade. Josten Børrelsen, of Indre Torhop, is thanked for providing boats for both the dry run and the final excursion (2007/2008). The Larsen brothers—Reidar, Jan, and Trygve—of Vestertana and Asbjørn Andersen of Berlevåg (see back cover of book) are collectively thanked for their willingness to take the participants in their boats to Stáhpogieddi (Stappogiedde) and northward, at extremely short notice in 2008. Grasbakken Hut and Boat Hire are thanked for their expertise with the boat trips to Skjåholmen and Vieranjar'ga in both 2007 and 2008. Mandy Hofmann kindly provided some extra global positioning system data taken during the trip, and Tony Spencer made many useful comments on the original manuscript, as well as allowing us to use his photograph for Figure 44B. We thank Michael Wagreich and Jörn Peckmann for discussions on rates of diagenesis in carbonate rocks and Bernhard Grasemann for discussions on structures (all at the University of Vienna). Edwards thanks Statoil for funding his travel for both the dry run (2007) and field trip (2008). Rice thanks the 33rd IGC (Oslo) for similar funding, and both Christa and Rhian Hofmann for support in the field over the last 20 years.

Finally, we thank all the participants of the original trip for their enthusiasm and comments; Michael Houmark-Nielsen, Paul Hoffman, Camille Partin, Tony Spencer, Mandy Hofmann, Erwan Le Guerroué, Julius Sovetov, Mark Smith, Noah Planavsky, Andrey Bekker, Guy Narbonne, and the entire Bourrouilh family.

Paul Hoffman and an anonymous reviewer are thanked for their comments on the first manuscript submitted. Bryan Hibbard, April Leo, and Kent Condie at the Geological Society of America are thanked for their help throughout the production of this guide. Talisman Energy Norge AS is thanked for generously paying for the color printing.

REFERENCES CITED

Arnaud, E., 2008, Deformation in the Neoproterozoic Smalfjord Formation, northern Norway: An indicator of glacial depositional conditions?: Sedimentology, v. 55, p. 335–356, doi:10.1111/j.1365-3091.2007.00903.x.

Arnaud, E., and Eyles, C.H., 2002, Glacial influence on Neoproterozoic sedimentation: The Smalfjord Formation, northern Norway: Sedimentology, v. 49, p. 765–788, doi:10.1046/j.1365-3091.2002.00466.x.

Arnaud, E., and Eyles, C.H., 2004, Glacial influence on Neoproterozoic sedimentation: The Smalfjord Formation, northern Norway: Reply: Sedimentology, v. 51, p. 1423–1430.

Arnaud, E., Halverson, G.P., and Shields-Zhou, G., 2011, The Geological Record of Neoproterozoic Glaciations: Geological Society of London Memoir 36, 735 p.

Baarli, B.G., Levine, R., and Johnson, M.E., 2006, The Late Neoproterozoic Smalfjord Formation of the Varanger Peninsula in northern Norway: A shallow fjord deposit: Norwegian Journal of Geology, v. 86, p. 133–150.

Banks, N.L., 1970, Trace fossils from the Late Precambrian and Lower Cambrian of Finnmark, North Norway, in Crimes, T.P., and Harper, J.C., eds., Trace Fossils: Geological Journal, v. 3, p. 19–34.

Banks, N.L., 1973, Innerelv Member: Late Precambrian marine shelf deposit, East Finnmark: Norges Geologiske Undersøkelse, v. 288, p. 7–25.

Banks, N.L., and Røe, S.-L., 1974, Sedimentology of the Late Precambrian Golneselv Formation, Varangerfjorden, Finnmark: Norges Geologiske Undersøkelse, v. 303, p. 17–38.

Banks, N.L., Edwards, M.B., Geddes, W.P., Hobday, D.K., and Reading, H.G., 1971, Late Precambrian and Cambro-Ordovician sedimentation in East Finnmark: Norges Geologiske Undersøkelse, v. 269, p. 197–236.

Banks, N.L., Hobday, D.K., Reading, H.G., and Taylor, P.N., 1974, Stratigraphy of the Late Precambrian 'Older Sandstone Series' of the Varangerfjord area, Finnmark: Norges Geologisk Undersøkelse, v. 303, p. 1–15.

Beckinsale, R.D., Reading, H.G., and Rex, D., 1976, Potassium-argon ages for basic dykes from East Finnmark: Stratigraphic and structural implications: Scottish Journal of Geology, v. 12, p. 51–65, doi:10.1144/sjg12010051.

Bestmann, M., Rice, A.H.N., Langenhorst, F., Grasemann, B., and Heidelbach, F., 2006, Subglacial bedrock welding associated with glacial earthquakes: Journal of the Geological Society of London, v. 163, p. 417–420, doi:10.1144/0016-764920-164.

Bingen, B., Griffin, W.L., Torsvik, T.H., and Saeed, A., 2005, Timing of Late Neoproterozoic glaciation on Baltica constrained by detrital zircon geochronology in the Hedmark Group, south-west Norway: Terra Nova, v. 17, p. 250–258, doi:10.1111/j.1365-3121.2005.00609.x.

Bjørlykke, K.O., 1967, The Eocambrian 'Reusch moraine' at Bigganjargga and the geology around Varangerfjord, northern Norway: Norges Geologiske Undersøkelse, v. 251, p. 18–44.

Bowring, S., Myrow, P., Landing, E., Ramezani, J., and Grotzinger, J., 2003, Geochronological constraints on terminal Neoproterozoic events and the rise of metazoans: Geophysical Research Abstracts, v. 5, no. 13, p. 219.

Bowring, S.A., Grotzinger, J.P., Condon, D.J., Ramezani, J., Newall, M.J., and Allen, P.A., 2007, Geochronologic constraints on the chronostratigraphic framework of the Neoproterozoic Huqf Supergroup, Sultanate of Oman: American Journal of Science, v. 307, p. 1097–1145, doi:10.2475/10.2007.01.

Cawood, P.A., and Piarevsky, S.A., 2006, Was Baltica right-way-up or upsidedown in the Neoproterozoic?: Journal of the Geological Society of London, v. 163, p. 753–759, doi:10.1144/0016-76492005-126.

Chapman, T.J., Gayer, R.A., and Williams, G.D., 1985, Structural cross-sections through the Finnmark Caledonides and timing of the Finnmarkian event, in Gee, D.G., and Sturt, B.A., eds., The Caledonide Orogen—Scandinavia and Related Areas: Chichester, UK, John Wiley & Sons, p. 593–610.

Cocks, L.R.M., and Torsvik, T.H., 2005, Baltica from the Late Precambrian to mid-Palaeozoic times: The gain and loss of a terrane's identity: Earth-Science Reviews, v. 72, p. 39–66, doi:10.1016/j.earscirev.2005.04.001.

Condon, D.J., and Bowring, S., 2011, A user's guide to Neoproterozoic geochronology, in Arnaud, E., Halverson, G.P, and Shields-Zhou, G., eds., The Geological Record of Neoproterozoic Glaciations: Geological Society of London Memoir 36, p. 135–150.

Crimes, T.P., and McIlroy, D., 1999, A biota of Ediacaran aspect from the Lower Cambrian strata on the Digermul Peninsula, Arctic Norway: Geological Magazine, v. 136, p. 633–642, doi:10.1017/S0016756899003179.

Crowell, J.C., 1964, Climactic significance of sedimentary deposits containing dispersed megaclasts, in Nairn, A.E.M., ed., Problems in Palaeoclimatology: London, Interscience Publishers, p. 86–99.

Crowell, J.C., 1999, Pre-Mesozoic Ice Ages: Their Bearing on Understanding the Climate System: Geological Society of America Memoir 192, 106 p.

Dal, A., 1900, Geologiske iagttagelser omkring Varangerfjorden: Norges Geologiske Undersøkelse, v. 28, p. 1–16.

Edwards, M.B., 1972, Glacial, Interglacial and Postglacial Sedimentation in a Late Precambrian Shelf Environment [Ph.D. thesis]: Oxford, UK, Oxford University, 284 p. (Free download available at http://www.marcedwards.com.)

Edwards, M.B., 1975, Glacial retreat sedimentation in the Smalfjord Formation, Late Precambrian, North Norway: Sedimentology, v. 22, p. 75–94, doi:10.1111/j.1365-3091.1975.tb00284.x.

Edwards, M.B., 1979, Late Precambrian glacial loessites from North Norway and Svalbard: Journal of Sedimentary Petrology, v. 49, p. 85–92.

Edwards, M.B., 1984, Sedimentology of the Upper Proterozoic glacial record, Vestertana Group, Finnmark, North Norway: Norges Geologiske Undersøkelse Bulletin, v. 394, p. 1–76.

Edwards, M.B., 1997, Discussion of glacial or non-glacial origin for the Bigganjargga tillite, Finnmark, northern Norway: Geological Magazine, v. 134, p. 873–876.

Edwards, M.B., 2004, Glacial influence on Neoproterozoic sedimentation: The Smalfjord Formation, northern Norway: Discussion: Sedimentology, v. 51, p. 1409–1417, doi:10.1111/j.1365-3091.2004.00674.x.

Fairchild, I.J., and Kennedy, M.J., 2007, Neoproterozoic glaciation in the Earth system: Journal of the Geological Society of London, v. 164, p. 895–921, doi:10.1144/0016-76492006-191.

Fanning, C.M., Ludwig, K.R., Forbes, B.G., and Preiss, W.V., 1986, Single and multiple grain U-Pb zircon analyses for the early Adelaidean Rook Tuff, Willouran Ranges, South Australia: Geological Society of Australia: Abstracts, v. 15, p. 71–72.

Farmer, J., Vidal, G., Moczydłowska, A., Strauss, H., Ahlberg, P., and Siedlecka, A., 1992, Ediacaran fossils from the Innerelv Member (Late Proterozoic) of the Tanafjorden area, northeastern Finnmark: Geological Magazine, v. 129, p. 181–195, doi:10.1017/S001675680000827X.

Føyn, S., 1937, The Eo-cambrian Series of the Tana District, northern Norway: Norsk Geologisk Tidsskrift, v. 17, p. 65–164.

Føyn, S., 1960, Tanafjord to Laksefjord, in Dons, J.A., ed., 21st International Geological Congress, Guide to Excursion A3: Norges Geologisk Undersøkelse, v. 212a, p. 45–57.

Føyn, S., and Glaessner, M.F., 1979, Platysolenites, other animal fossils, and the Precambrian-Cambrian transition in Norway: Norsk Geologisk Tidsskrift, v. 59, p. 25–46.

Føyn, S., and Siedlecki, S., 1980, Glacial stadials and interstadials in the Late Precambrian Smalfjord tillite on Laksefjordvidda, Finnmark, North Norway: Norges Geologiske Undersøkelse, v. 358, p. 31–45.

Gaertner, H.R. von, 1943, Bemerkungen über den Tillit von Bigganjargga am Varangerfjord: Geologische Rundschau, v. 34, p. 226–231, doi:10.1007/BF01766413.

Gayer, R.A., and Rice, A.H.N., 1989, Palaeogeographic reconstruction of the pre- to syn-Iapetus rifting sediments in the Caledonides of Finnmark, N. Norway, in Gayer, R.A., ed., The Caledonide Geology of Scandinavia: London, Graham & Trotman, p. 127–139.

Gayer, R.A., Rice, A.H.N., Roberts, D., Townsend, C., and Welbon, A., 1987, Restoration of the Caledonian Baltoscandian margin from balanced cross-sections: The problem of excess continental crust: Transactions of the Royal Society of Edinburgh–Earth Sciences, v. 78, p. 197–217, doi:10.1017/S026359330001110X.

Grey, K., Hill, A.C., and Calver, C., 2011, Biostratigraphy and stratigraphic subdivision of Cryogenian successions of Australia in a global context, in Arnaud, E., Halverson, G.P., and Shields-Zhou, G., eds., The Geological Record of Neoproterozoic Glaciations: Geological Society of London Memoir 36, p. 113–134.

Guise, P.G., and Roberts, D., 2002, Devonian ages from $^{40}Ar/^{39}Ar$ dating of plagioclase in dolerite dykes, eastern Varanger Peninsula, North Norway: Norges Geologiske Undersøkelse Bulletin, v. 440, p. 27–37.

Halverson, G.P., and Shields-Zhou, G., 2011, Chemostratigraphy and the Neoproterozoic glaciations, in Arnaud, E., Halverson, G.P., and Shields-Zhou, G., eds., The Geological Record of Neoproterozoic Glaciations: Geological Society of London Memoir 36, p. 51–66.

Halverson, G.P., Hoffman, P.F., Schrag, D.P., Maloof, A.C., and Rice, A.H.N., 2005, Towards a Neoproterozoic composite carbon isotope record: Geological Society of America Bulletin, v. 117, p. 1181–1207, doi:10.1130/B25630.1.

Hansen, T.A., 1992, Sedimentologiske og Stratigrafiske Undersøkelser av den Sen-Prekambriske Smalfjord Formasjonen, Øst-Finnmark [Cand. Real. thesis]: University of Troms, 58 p.

Harland, W.B., 1964, Evidence for a Late Precambrian glaciation and its significance, in Nairn, A.E.M., ed., Problems in Palaeoclimatology: London, John Wiley & Sons, p. 119–149.

Hartz, E.H., and Torsvik, T.H., 2002, Baltica upside down: A new plate tectonic model for Rodinia and the Iapetus Ocean: Geology, v. 30, p. 255–258, doi:10.1130/0091-7613(2002)030<0255:BUDANP>2.0.CO;2.

Hobday, D.K., 1974, Interaction between fluvial and marine processes in the lower part of the Late Precambrian Vadsø Group, Finnmark: Norges Geologiske Undersøkelse, v. 303, p. 39–58.

Hoffmann, K.-H., Condon, D.J., Bowring, S.A., and Crowley, J.L., 2004, U-Pb zircon date from the Neoproterozoic Ghaub Formation, Namibia: Constraints on Marinoan glaciation: Geology, v. 32, p. 817–820, doi:10.1130/G20519.1.

Hofmann, P.F., 2011, A history of Neoproterozoic glacial geology, 1871–1997, in Arnaud, E., Halverson, G.P., and Shields-Zhou, G., eds., The Geological Record of Neoproterozoic Glaciations: Geological Society of London Memoir 36, p. 17–37.

Holtedahl, O., 1918, Contribution to the geology of Finnmarken: Norges Geologiske Undersøkelse, v. 84, p. 1–314 (in Norwegian with English summary).

Hossack, J.R., and Cooper, M.A., 1986, Collision tectonics in the Scandinavian Caledonides, in Coward, M.P., and Ries, A.C., eds., Collision Tectonics: Geological Society of London Special Publication 19, p. 285–304.

Jensen, P.A., and Wulff-Pedersen, E., 1996, Glacial or non-glacial origin for the Bigganjarga tillite, Finnmark, northern Norway?: Geological Magazine, v. 133, p. 137–145, doi:10.1017/S0016756800008657.

Johnson, H.D., 1975, Tide- and wave-dominated inshore and shoreline sequences from the Late Precambrian, Finnmark, North Norway: Sedimentology, v. 22, p. 45–74, doi:10.1111/j.1365-3091.1975.tb00283.x.

Johnson, H.D., 1978, Facies distributions and lithostratigraphic correlation in the Late Precambrian Ekkerøy Formation, east Finnmark, Norway: Norsk Geologisk Tidsskrift, v. 58, p. 175–190.

Johnson, H.D., Levell, B.K., and Siedlecki, S., 1978, Late Precambrian sedimentary rocks in East Finnmark, North Norway, and their relationship to the Trollfjord-Komagelva fault: Journal of the Geological Society of London, v. 135, p. 517–533, doi:10.1144/gsjgs.135.5.0517.

Kennedy, M.J., Runnegar, B., Prave, A.R., Hoffman, K.H., and Arthur, M.A., 1998, Two or four Neoproterozoic glaciations?: Geology, v. 26, p. 1059–1063, doi:10.1130/0091-7613(1998)026<1059:TOFNG>2.3.CO;2.

Laajoki, K., 2001, Additional observations on the late Proterozoic Varangerfjorden unconformity, Finnmark, northern Norway: Bulletin of the Geological Survey of Finland, v. 73, p. 17–34.

Laajoki, K., 2002, New evidence of glacial abrasion of the Late Proterozoic unconformity around Varangerfjorden, northern Norway, in Altermann, W., and Corcoran, P.L., eds., Precambrian Sedimentary Environments: A

Modern Approach to Ancient Depositional Systems: International Association of Sedimentologists Special Publication 33, p. 405–436.

Laajoki, K., 2003, The Larajæg'gi outcrop—A large combined Neoproterozoic/Pleistocene roche moutonnée at Karlebotn, Finnmark, North Norway: Norwegian Journal of Geology, v. 84, p. 107–115.

Le Guerroué, E., Allen, P.A., and Cozzi, A., 2006, Chemostratigraphic and sedimentological framework of the largest negative carbon isotopic excursion in Earth history: The Neoproterozoic Shuram Formation (Nafun Group, Oman): Precambrian Research, v. 146, p. 68–92, doi:10.1016/j.precamres.2006.01.007.

Macdonald, F.A., Schmitz, M.D., Crowley, J.L., Roots, C.F., Jones, D.S., Maloof, A.C., Strauss, J.V., Cohen, P.A., Johnston, D.T., and Schrag, D.P., 2010, Calibrating the Cryogenian: Science, v. 327, p. 1241–1243, doi:10.1126/science.1183325.

McCay, G.A., Prave, A.R., Alsop, G.I., and Fallick, A.E., 2006, Glacial trinity: Neoproterozoic Earth history within the British-Irish Caledonides: Geology, v. 34, p. 909–912, doi:10.1130/G22694A.1.

Pickering, K.T., 1981, The Kongsfjord Formation—A Late Precambrian submarine fan in north-east Finnmark, Norway: Norges Geologiske Undersøkelse, v. 367, p. 77–104.

Prave, A.R., Fallick, A.E., Thomas, C.W., and Graham, C.M., 2009, A composite C-isotope profile for the Neoproterozoic Dalradian Supergroup of Scotland and Ireland: Journal of the Geological Society of London, v. 166, p. 845–857, doi:10.1144/0016-76492008-131.

Reading, H.G., 1965, Eocambrian and Lower Palaeozoic geology of the Digermul Peninsula, Tanafjord, Finnmark: Norges Geologiske Undersøkelse, v. 234, p.167–191.

Reading, H.G., and Walker, R.G., 1966, Sedimentation of Eocambrian tillites and associated sediments in Finnmark, North Norway: Palaeogeography, Palaeoclimatology, Palaeoecology, v. 2, p. 177–212.

Reusch, H., 1891, Skuringsmærker og morængrus eftervist i Finnmarken fra en periode meget ældre end "istiden": Norges Geologiske Undersøkelse, v. 1, p. 78–85 (English summary p. 97–100).

Rice, A.H.N., 1994, Stratigraphic overlap of the late Proterozoic Vadsø and Barents Sea Groups and correlation across the Trollfjorden-Komagelva fault, Finnmark, North Norway: Norsk Geologisk Tidsskrift, v. 74, p. 48–57.

Rice, A.H.N., 2001, Field evidence for thrusting of the basement rocks coring tectonic windows in the Scandinavian Caledonides; an insight from the Kunes Nappe, Finnmark, Norway: Norwegian Journal of Geology, v. 81, p. 321–328.

Rice, A.H.N., and Frank, W., 2003, Early Caledonian (Finnmarkian) events reassessed in Finnmark: $^{40}Ar/^{39}Ar$ cleavage age data from NW Varangerhalvøya, N. Norway: Tectonophysics, v. 374, p. 219–236, doi:10.1016/S0040-1951(03)00240-3.

Rice, A.H.N., and Hofmann, Ch.-Ch., 2000, Evidence for a glacial origin of the Neoproterozoic III striations at Oaibaččannjar'ga, Finnmark, northern Norway: Geological Magazine, v. 137, p. 355–366, doi:10.1017/S0016756800004222.

Rice, A.H.N., and Hofmann, Ch.-Ch., 2001, The transition from Neoproterozoic glacial to interglacial sedimentation near Hammarnes, East Finnmark, North Norway: Norwegian Journal of Geology, v. 81, p. 257–262.

Rice, A.H.N., and Reiz, W., 1994, The structural relations and regional tectonic implications of metadolerite dykes in the Kongsfjord Formation, North Varanger region, Finnmark, N. Norway: Norsk Geologisk Tidsskrift, v. 74, p. 152–165.

Rice, A.H.N., and Townsend, C., 1996, Correlation of the late Precambrian Ekkerøya Formation (Vadsø Group; E. Finnmark) and the Brennelvfjord Interbedded Member (Porsangerfjord Group; W. Finnmark), N. Norwegian Caledonides: Norsk Geologisk Tidsskrift, v. 76, p. 55–61.

Rice, A.H.N., Gayer, R.A., Robinson, D., and Bevins, R.E., 1989a, Strike-slip restoration of the Barents Sea Caledonides terrane, Finnmark, North Norway: Tectonics, v. 8, p. 247–264, doi:10.1029/TC008i002p00247.

Rice, A.H.N., Bevins, R.E., Robinson, D., and Roberts, D., 1989b, Evolution of low-grade metamorphic zones in the Caledonides of Finnmark, N. Norway, in Gayer, R.A., ed., The Caledonide Geology of Scandinavia: London, Graham & Trotman, p. 177–191.

Rice, A.H.N., Hofmann, Ch.-Ch., and Pettersen, A., 2001, A new sedimentary succession from the southern margin of the Neoproterozoic Gaissa Basin, south Varangerfjord, North Norway; the Lattanjar'ga unit: Norsk Geologisk Tidsskrift, v. 81, p. 41–48.

Rice, A.H.N., Ntaflos, T., Gayer, R.A., and Beckinsale, R.D., 2004, Metadolerite geochronology and dolerite geochemistry from East Finnmark, northern Scandinavian Caledonides: Geological Magazine, v. 141, p. 301–318, doi:10.1017/S001675680300788X.

Rice, A.H.N., Edwards, M.B., Hansen, T.A., Arnaud, E., and Halverson, G.P., 2011, Glaciogenic rocks of the Neoproterozoic Smalfjord and Mortensnes Formations, Vestertana Group, E. Finnmark, Norway, in Arnaud, E., Halverson, G.P., and Shields-Zhou, G., eds., The Geological Record of Neoproterozoic Glaciations: Geological Society of London Memoir 36, p. 593–602.

Roberts, D., 1972, Tectonic deformation in the Barents Sea Region of Varanger Peninsula, Finnmark: Norges geologiske undersøkelse, v. 282, p. 1–39.

Roberts, D., 1975, Geochemistry of dolerite and metadolerite dykes from Varanger Peninsula, Finnmark, North Norway: Norges Geologiske Undersøkelse, v. 322, p. 55–72.

Roberts, D., 2003, Shipwrecked on Top of the World: Four against the Arctic: London, Little Brown, 320 p.

Roberts, D., and Siedlecka, A., 2002, Timanian orogenic deformation along the northeastern margin of Baltica, northwest Russia and northeast Norway, and Avalonian-Cadomian connections: Tectonophysics, v. 352, p. 169–184, doi:10.1016/S0040-1951(02)00195-6.

Røe, S.-L., 1970, Correlation between the late Precambrian Olders Sandstone Series of the Varangerfjord and Tanafjord areas: Norges Geologiske Undersøkelse, v. 266, p. 230–245.

Røe, S.-L., 1975, Stratification, Sedimentary Processes and Depositional Environments of Part of the Late Precambrian Vadsø Group, Varangerfjord Area, Finnmark [Cand. Real thesis]: Bergen, Norway, University of Bergen, 187 p.

Røe, S.-L., 1987, Cross-strata and bedforms of probable transitional dune to upper stage plane-bed origin from a Late Precambrian fluvial sandstone, northern Norway: Sedimentology, v. 34, p. 89–101, doi:10.1111/j.1365-3091.1987.tb00562.x.

Røe, S.-L., 2003, Neoproterozoic peripheral-basin deposits in eastern Finnmark, N. Norway: Stratigraphic revision and palaeotectonic implications: Norwegian Journal of Geology, v. 83, p. 259–274.

Rooney, A.D., Chew, D.M., and Selby, D., 2011, Re-Os geochronology of the Neoproterozoic-Cambrian Dalradian Supergroup of Scotland and Ireland: Implications for Neoproterozoic stratigraphy, glaciations and Re-Os systematic: Precambrian Research, v. 185, p. 202–214.

Rosendahl, H., 1931, Bidrag til Varangernesets geologi: Norsk Geologisk Tidsskrift, v. 12, p. 487–506.

Rosendahl, H., 1945, Prekambrium-Eokambrium I Finnmark: Norsk Geologisk Tidsskrift, v. 25, p. 327–349.

Schermerhorn, L.J.G., 1974, Late Precambrian mixtites: Glacial or nonglacial: American Journal of Science, v. 274, p. 673–824, doi:10.2475/ajs.274.7.673.

Schiøtz, O.E., 1898, Om Dr Reusch's präglaciale skuringdmerker: Nyt mag: Naturv, v. 36, p. 1–10.

Shields, G.A., 2002, A chemical explanation for the mid-Neoproterozoic disappearance of molar-tooth structure ~750 Ma: Terra Nova, v. 14, p. 108–113, doi:10.1046/j.1365-3121.2002.00396.x.

Siedlecka, A., 1972, Konsgfjord Formation—A Late Precambrian flysch sequence from the Varanger Peninsula, Finnmark: Norges Geologiske Undersøkelse, v. 278, p. 41–80.

Siedlecka, A., 1975, Late Precambrian stratigraphy and structure of the northeastern margin of the Fennoscandian Shield (East Finnmark-Timan region): Norges Geologiske Undersøkelse, v. 336, p. 313–348.

Siedlecka, A., 1978, Late Precambrian tidal-flat deposits and algal stromatolites in the Båtsfjord Formation, East Finnmark, North Norway: Sedimentary Geology, v. 21, p. 277–310, doi:10.1016/0037-0738(78)90023-4.

Siedlecka, A., 1982, Supralittoral ponded algal stromatolites of the Late Precambrian Annijåkka Member of the Båtsfjord Formation, Varanger Peninsula, North Norway: Precambrian Research, v. 18, p. 319–345, doi:10.1016/0301-9268(82)90007-9.

Siedlecka, A., 1985, Development of the Upper Proterozoic sedimentary basins of the Varanger Peninsula, East Finnmark: Geological Survey of Finland Bulletin, v. 331, p. 175–185.

Siedlecka, A., compiler, 1990, Varangerbotn Berggrunnskart 2335 3, Foreløpig Utgave: Norges Geologisk Undersøkelse; scale 1:50,000, 1 sheet.

Siedlecka, A., 1995, Neoproterozoic sedimentation on the Rybachi and Sredni Peninsulas and Kildin Island, NW Kola, Russia: Norges Geologiske Undersøkelse Bulletin, v. 427, p. 52–55.

Siedlecka, A., and Edwards, M.B., 1980, Lithostratigraphy and sedimentology of the Riphean Båsnæring Formation, Varanger Peninsula, North Norway: Norges Geologisk Undersøkelse, v. 355, p. 27–47.

Siedlecka, A., and Roberts, D., 1992, The Bedrock Geology of Varanger Peninsula, Finnmark, North Norway: An Excursion Guide: Norges Geologisk Undersøkelse Special Publication 5, 45 p.

Siedlecka, A., and Siedlecki, S., 1971, Late Precambrian sedimentary rocks of the Tanafjord-Varangerfjord region of Varanger Peninsula, northern Norway: Norges Geologiske Undersøkelse, v. 269, p. 246–294.

Siedlecka, A., Pickering, K.T., and Edwards, M.B., 1989, Upper Proterozoic passive margin delta complex, Finnmark, North Norway, in Whateley, M.K.G., and Pickering, K.T., eds., Deltas: Sites and Traps for Fossil Fuels: Geological Society of London Special Publication 41, p. 205–219.

Siedlecka, A., Roberts, D., Nystuen, J.P., and Olovyanishnikov, V.G., 2004, Northeastern and northwestern margins of Baltica in Neoproterozoic time: Evidence from the Timanian and Caledonian orogens: Geological Society of London Memoir 30, p. 169–190.

Siedlecki, S., compiler, 1980, Geologisk Kart over Norge, Berggrunnskart Vadsø: Trondheim, Norway, Norges Geologiske Undersøkelse, scale 1:250,000, 1 sheet.

Spjeldnaes, N., 1964, The Eocambrian glaciation in Norway: Geologische Rundschau, v. 54, p. 24–45, doi:10.1007/BF01821168.

Stodt, F., 1987, Sedimentologie, Spurenfossilien und Weichkörper-Metazoan der Dividal Gruppe (Wendium-Unterkambrium) im Tornetråskgebiet/ Nordschweden [Ph.D. thesis]: Marburg, Germany, Phillip University of Marburg, 119 p.

Stodt, F., Rice, A.H.N., Björklund, L., Bax, G., Halverson, G.P., and Pharaoh, T.C., 2011, Evidence of late Neoproterozoic glaciation in the Caledonides of NW Scandinavia, in Arnaud, E., Halverson, G.P., and Shields-Zhou, G., eds., The Geological Record of Neoproterozoic Glaciations: Geological Society of London Memoir 36, p. 603–612.

Strahan, A., 1897, On glacial phenomena of Palaeozoic age in the Varanger Fjord: Quarterly Journal of the Geological Society, v. 53, p. 137–146, doi:10.1144/GSL.JGS.1897.053.01-04.11.

Townsend, C., 1986, Thrust Tectonics within the Caledonides of Northern Norway [Ph.D. thesis]: Cardiff, UK, University of Wales, 210 p.

Townsend, C., Roberts, D., Rice, A.H.N., and Gayer, R.A., 1986, The Gaissa nappe, Finnmark, North Norway: An example of a deeply eroded external imbricate zone within the Scandinavian Caledonides: Journal of Structural Geology, v. 8, p. 431–440, doi:10.1016/0191-8141(86)90061-1.

Townsend, C., Rice, A.H.N., and Mackay, A., 1989, The structure and stratigraphy of the southwestern portion of the Gaissa thrust belt and adjacent Kalak nappe complex, Finnmark, N Norway, in Gayer, R.A., ed., The Caledonide Geology of Scandinavia: London, Graham & Trotman, p. 111–126.

Vidal, G., 1981, Micropaleontology and biostratigraphy of the Upper Proterozoic and Lower Cambrian sequences in East Finnmark, northern Norway: Norges Geologiske Undersøkelse, v. 362, p. 1–53.

Vidal, G., and Siedlecka, A., 1983, Planktonic, acid-resistant microfossils from the Upper Proterozoic strata of the Barents Sea region of Varanger Peninsula, East Finnmark, northern Norway: Norges Geologiske Undersøkelse, v. 382, p. 45–79.

Williams, D.M., 1976, A revised stratigraphy of the Gaissa nappe, Finnmark: Norges Geologiske Undersøkelse, v. 324, p. 63–78.

Williams, D.M., 1979, Structural development of the Gaissa nappe in the Finnmark Caledonides, North Norway: Norges Geologiske Undersøkelse, v. 348, p. 93–104.

Zhang, S., Jiang, G., and Han, Y., 2008, The age of the Nantuo Formation and the Nantuo glaciation in South China: Terra Nova, v. 20, p. 289–294, doi:10.1111/j.1365-3121.2008.00819.x.

MANUSCRIPT ACCEPTED BY THE SOCIETY 27 DECEMBER 2011